P9-CQP-055

SAFETY OF CHEMICALS IN FOOD
Chemical Contaminants

ELLIS HORWOOD SERIES IN FOOD SCIENCE, MANAGEMENT AND TECHNOLOGY

Editor-in-Chief:
I. D. MORTON, Emeritus Professor and formerly Head of Department of Food and Nutritional Science, King's College, University of London.
Series Editors:
D. H. WATSON, Ministry of Agriculture, Fisheries and Food, and M. J. LEWIS, Department of Food Science and Technology, University of Reading

Ali and Robinson	**TRADITIONAL FERMENTED FOODS**
Bourgeois, Mescle and Zucca	**FOOD MICROBIOLOGY: Volumes 1 and 2**
Cheng	**FOOD MACHINERY: For the Production of Cereal Foods, Snack Foods and Confectionery**
Dennis and Stringer	**CHILLED FOODS: A Comprehensive Guide**
Fellows	**FOOD PROCESSING TECHNOLOGY: Principles and Practice**
Girard	**TECHNOLOGY OF MEAT AND MEAT PRODUCTS**
Grandison and Lewis	**SEPARATION PROCESSES: Principles and Applications**
Grandison, Lewis and Wilbey	**DAIRY TECHNOLOGY**
Henderson and Grootveld	**FOOD IRRADIATION: Molecular and Medical Implications**
Hill	**EPIDEMIOLOGY OF DIET AND CANCER**
Lewis	**PRACTICAL PROPERTIES OF FOOD AND FOOD PROCESSING SYSTEMS**
Morton and Lenges	**EDUCATION AND TRAINING IN FOOD SCIENCE: A Changing Scene**
Niranjan and de Alwis	**FOOD PROCESS MODELLING: Relating Food Process Design, Food Safety and Quality**
Watson	**SAFETY OF CHEMICALS IN FOOD: Chemical Contaminants**
Watson	**FOOD SCIENCE REVIEWS: Volume 1 Food Hygiene and Safety**
Wiese, Jackson, Wiese and Davis	**FOOD CONTAINER CORROSION**

SAFETY OF CHEMICALS IN FOOD
Chemical Contaminants

Editor:
D. H. WATSON
Ministry of Agriculture, Fisheries and Food, London

ELLIS HORWOOD
NEW YORK LONDON TORONTO SYDNEY TOKYO SINGAPORE

First published in 1993 by
Ellis Horwood Limited
Market Cross House, Cooper Street,
Chichester, West Sussex, PO19 1EB, England
A division of
Simon & Schuster International Group

Printed and bound in Great Britain
by Hartnolls, Bodmin

British Library Cataloguing in Publication Data

A catalogue record for this book is available from the British Library

ISBN 0–13–787862–1

Library of Congress Cataloging-in-Publication Data

Available from the publisher

Table of contents

Preface

Although there are many books that review particular groups of chemical contaminants in food in great depth, there is no general guide that can help introduce food scientists, and others, to this important area of work. There is also much media coverage, some of it inaccurate, but there is not a general source of reference on the extensive but still developing subject of chemical contaminants in food. The purpose of writing this book is to give a general introduction to the main groups of these chemicals—to provide the reader with information about which chemicals are involved, why they contaminate food and how their impact on the food chain can be measured. The major groups of chemical contaminants are reviewed in Chapters 2 to 8. There is not a chapter on radioactive elements since scientific work on these substances is quite distinct—it is the contamination of food with radioactivity that is of primary interest whereas for other chemical contaminants it is the chemicals themselves that are mainly studied.

This book goes into some detail about the methodology used in scientific work on chemical contaminants in food. Methods of analysing food have been reviewed elsewhere (see for example *Analysis of food contaminants* edited by Dr John Gilbert, publ. Elsevier Applied Science, 1984), but the general principles and approaches of analytical work are summarized here where it is essential for a proper understanding of work on the chemicals. There is less information, and little guidance, in the literature about how food chemical surveillance is carried out, or on the fundamentals of estimating consumers' intakes of chemical contaminants. These two key areas of methodology are therefore described and reviewed in detail, in Chapters 9 and 10.

The study of chemical contaminants in food is still a relatively young science. Some key areas are in particular need of further work—these are reviewed in Chapter 11. It is hoped that the reader will be stimulated by this chapter, and the rest of the book, to find out more about chemical contamination of food. To help in this, references to the literature have been chosen not for the usual reason of providing experimental detail but to expand upon and illustrate further some of the main points. There is also a part of Chapter 9 (section 9.8) which introduces the reader to the extensive range of research journals and other sources of experimental information on surveillance for chemical contaminants in food.

This book should serve as a useful introduction to those with a professional interest in food safety, as well as those who have to learn about chemical contamination as part of their formal education. It provides considerable data and information, as well as insights into the thinking of scientists who work on chemical contamination in food. The opinions stated are those of the authors—they are not official statements of the organization for which the authors work. Much of the illustrative data and information are drawn from scientific work by the Government in the UK because its programme of work is extensive and generally at the forefront of this scientific subject. And the findings of UK work have been confirmed in other countries. But reference is made to work in other parts of the world to provide an international picture.

Last, but not least, I am happy to acknowledge the efforts of many colleagues over the years to help establish chemical contamination of food as a scientific subject, and I am also pleased to recognize the advice and constructive comments of Dr Richard Burt on the text of this book.

D. H. Watson
October 1992
R242, Ministry of Agriculture, Fisheries and Food,
Ergon House,
c/o Nobel House,
17 Smith Square,
London SW1P 3JR, UK.

Glossary of terms and abbreviations

ACP: Advisory Committee on Pesticides (see Fig. 9.2).

Acute toxicity: Toxicity that occurs soon after exposure to the toxic substance (within hours or a few days).

ADI: Acceptable Daily Intake. The amount of a chemical which can be consumed every day for an individual's entire lifetime in the practical certainty, on the basis of all the known facts, that no harm will result. The ADI is expressed as milligrams of the chemical per kilogram bodyweight of the consumer.

AQA: Analytical quality assurance.

CCRVDF: Codex Committee on Residues of Veterinary Drugs in Foods.

Carcinogen: A substance that causes cancer.

Chronic toxicity: Toxicity that occurs a considerable time after exposure to a toxic substance (usually months or longer).

Concentration (= level) of a chemical contaminant in food or drink: The amount of the chemical contaminant per unit weight of food or drink. Several different units are used:

mg/kg (equivalent to one part per 10^6) ng/kg (equivalent to one part per 10^{12})
μg/kg (equivalent to one part per 10^9) pg/kg (equivalent to one part per 10^{15}).

CVMP: EC Committee for Veterinary Medicinal Products.

EC: European Community.

FAO: Food and Agriculture Organization (of the United Nations).

FSIS: US Food Safety and Inspection Service.

GC: Gas chromatography.

HCB: Hexachlorobenzene.

HCH: Hexachlorocyclohexane.

HPLC: High performance (or high pressure) liquid chromatography.

Limit of detection: The lowest concentration of a chemical contaminant that can be identified and quantitatively measured in a specified food or drink, or raw material used in their production, with an acceptable degree of certainty by a given method of analysis.

MAFF: Ministry of Agriculture, Fisheries and Food (a UK Government department).

MRL: Maximum Residue Limit (for residues of pesticides and veterinary drugs in food). A definition for MRL as applied to work on veterinary residues is given in Chapter 2, section 2.2.4.1. A definition for MRL as applied to work on pesticides residues is as follows: the maximum concentration of pesticides residues likely to occur in or on a food commodity, either resulting from the use of the pesticide according to good agricultural practice (directly or indirectly for the production and/or protection of the commodity concerned) or arising from environmental sources, including former agricultural uses.
MS: Mass spectrometry.
Mycotoxins: Toxins produced by fungi.
NDMA: *N*-Nitrosodimethylamine.
NSS: UK State Veterinary Service's National Surveillance Scheme.
PCBs: Polychlorinated biphenyls.
PCDDs: Polychlorinated dibenzo-*p*-dioxins.
PCDFs: Polychlorinated dibenzofurans
SCF: Scientific Committee for Food (of the EC Commission).
SGCAFS: Steering Group on Chemical Aspects of Food Surveillance (see Fig. 9.2).
SI: Statutory Instrument (a type of legislation in the UK).
TDI: Tolerable Daily Intake (see ADI and p. 2).
TDS: UK Total Diet Study (see Chapter 10, section 10.4.2).
TEQ: Toxic equivalents.
VPC: Veterinary Products Committee (see Fig. 9.2, and Chapter 2, section 2.2.1).
WHO: World Health Organization (of the United Nations).

1

Introduction

D. H. Watson, Ministry of Agriculture, Fisheries and Food, R242, Ergon House, c/o Nobel House, 17 Smith Square, London SW1P 3JR, UK.

1.1 DEFINITION OF 'CHEMICAL CONTAMINANT'

The definition of a chemical contaminant used in this book is as follows:

> Any substance not intentionally added to food, which is present in such food as a result of the production (including operations carried out in crop husbandry, animal husbandry and veterinary medicine), manufacture, processing, preparation, treatment, packing, packaging, transport or holding of such food, or as a result of environmental contamination *or its production by a living organism.*

This is the definition used by the United Nations' Codex Committee on Food Additives and Contaminants, except for the last seven words in italics which are added here to include naturally occurring toxicants. Thus the definition covers all of the substances that are reviewed in this book. The chemicals can be classified into various categories according to why they are present in food (Table 1.1). It is worth noting that most chemical contaminants can arise from more than one activity. Indeed pesticides residues in food can result from nearly every activity listed in Table 1.1.

It should also be noted that 'food' in the above definition is generally taken to include drink by scientists who work on chemical contaminants. Both parts of the diet are included.

1.2 SCIENTIFIC WORK ON CHEMICAL CONTAMINANTS IN FOOD

The study of chemical contamination of food has developed as a scientific subject primarily over the last twenty years or so. Its relatively recent development has been due to a number of factors, of which the following are probably the main ones:

— Many methods of analysis have been developed which are suitable for detecting and quantifying the very low concentrations of chemicals in food. This has been part of a

Table 1.1. A classification of chemical contaminants in food

Activity	Related chemical contaminants that can occur in food
Crop production	Pesticides, nitrate, metals, naturally occurring toxicants
Animal production	Pesticides, veterinary drugs, metals
Food manufacture	Pesticides, metals, nitrate, nitrite, nitrosamines
Packaging of food	Chemicals migrating from packaging
Food storage	Pesticides, metals
Industrial	Environmental organic chemical contaminants (e.g. benzene), metals, pesticides

transfer of expertise and techniques from other areas of science to the study of food safety.

— The science of toxicology has developed rapidly making it possible to test the toxicity of a wide range of chemicals used by man, for example pesticides, veterinary drugs and industrial chemicals.

— It has been found that food can be chemically contaminated by several routes, from the use of agricultural and other chemicals.

— Growing public interest in food safety has been stimulated by media coverage of the results of some of this scientific work.

— An increasing variety of controls on chemical contamination of food has been introduced, both in individual countries and on international trade.

It is impossible to say which of these factors has been most important in stimulating the scientific work, but it is clear that the science has been allowed to develop quickly because of the development of analytical methods to detect chemical contaminants in food. This has been a key factor because it is the very low concentrations of chemical contaminants in food that distinguish them from most other forms of contamination of food. These concentrations are generally one milligram of chemical per kilogram of food (one part per million), or less. Levels of one part in 10^{14} are now detectable for some chemical contaminants in food (for example dioxins: Chapter 3). And it is essential that such extremely low concentrations can be detected because in many cases chemical contaminants occur only very rarely at greater concentrations in food. Not surprisingly it can be difficult to judge the toxicological significance of such low concentrations of chemicals in food. Nevertheless there has been considerable progress in identifying what the toxic effects might be, if any, and in deciding the maximum amounts of given chemical contaminants that can be consumed without risk to the consumer. The setting of tolerable daily intakes (TDIs) for chemical contaminants in food is becoming a key part of the study of these substances. In the past the term acceptable daily intake (ADI) was used but TDI is now preferred because exposure to chemical contaminants is tolerable rather than acceptable. These and other specialist terms used in this book are defined in the *Glossary of terms and abbreviations* at the front of the volume.

There are several areas of specialist methodology used in the study of chemical contaminants in food. Two of them—food surveillance and estimating consumer intakes of these substances—are central to work on one of the main themes of this book: the concentrations and incidences of chemical contaminants in food. These two methodological subjects are described in detail in Chapters 9 and 10 of this book. In summary the methods involved are as follows:

— *Food surveillance*: This is, in its broadest sense, keeping a watch on chemical contamination of the food supply. In practice surveillance is usually carried out by analysing samples of foods or raw materials (such as grain) for chemical contaminants. Surveys of food (and its raw materials) provide information about how often samples contain detectable contamination and what concentrations of contaminants are present. It is essential to have this information so that the next stage of investigation, estimating consumer intakes, can be carried out.
— *Estimating consumer intakes of chemical contaminants from the diet*: This uses the information provided by food surveillance, together with data on the amounts of food and drink that are consumed, to calculate intakes of the chemicals. There are also other ways of estimating such intakes, as described in Chapter 10.

These activities lead up to an estimate of how much of a given chemical contaminant is consumed. Comparing this with a TDI gives a straightforward measure of whether the chemical is a risk to consumers. This represents one step in the sequence of events in much of the current work on chemical contaminants in food:

— Identify the chemical concerned.
— Find out which types of food and drink are contaminated, and to what degree.
— Estimate dietary exposure to the contaminant.
— Judge the degree of risk that is involved.
— Take action to reduce the risk.
— Check that the action has been successful.

1.3 TYPES OF CHEMICAL CONTAMINANTS

Most chemical contaminants in food are relatively complex organic chemicals. Table 1.2 illustrates the variety of different chemicals involved. For each of the categories given in this table there are distinct sub-categories of chemicals. These can be classified as follows:

— by their chemical structures; for example environmental chemical contaminants can be classified as follows: dioxins, furans, other halogenated organic chemicals, polynuclear aromatic hydrocarbons, and other organic environmental chemicals;
— or by their usage; for example for pesticides the groups would include: insecticides, fungicides, herbicides, acaricides;
— or by their origins; for example for naturally occurring toxicants there are several categories: bacterial toxins, algal toxins, fungal toxins (mycotoxins) and higher plant toxicants.

Table 1.2. Different organic chemical contaminants

Substance	Category	Structure
Tetracycline	Veterinary drug	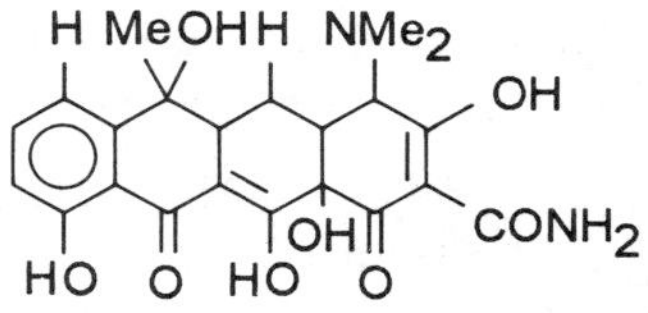 H MeOH H NMe_2 OH $CONH_2$ HO O HO OH O
p-Dichlorobenzene	Environmental contaminant	 Cl Cl
N-Nitrosodimethylamine	An *N*-nitrosamine	Me Me N.NO
Deoxynivalenol	Naturally occurring toxicant	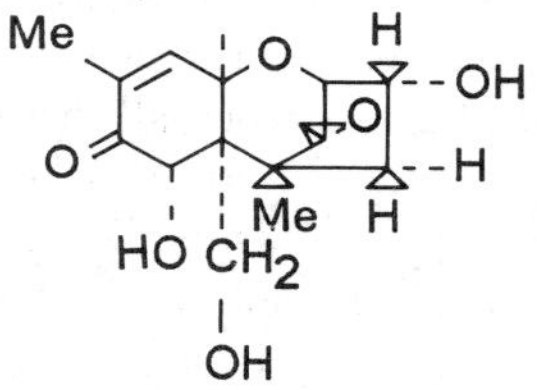 Me O H OH O H O Me H HO CH_2 OH
Styrene	Chemical in packaging	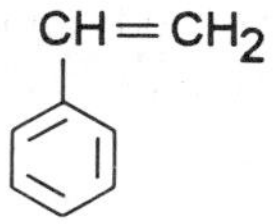 $CH{=}CH_2$
Dichlorvos	Pesticide	O ‖ $Cl_2C{=}CHOP(OCH_3)_2$

By convention veterinary drugs, pesticides and chemicals used in packaging are categorized by their uses, whilst the other chemicals that are discussed in this book are usually classified by their chemical properties. The one exception is naturally occurring toxicants which are grouped according to the organisms that produce them.

The numbers of chemicals vary considerably between the categories (Table 1.3). The largest category is industrial chemicals, although it is suspected that only a very small minority of the many chemicals that are used by industry (and hence can be present in the environment) will actually contaminate food. The next largest categories—natural

Table 1.3. Theoretical maximum numbers of chemicals that might contaminate food and numbers found in food

Category	Theoretical maximum numbers of chemicals[†]	Numbers of chemicals detected in food[†]
Veterinary drugs	< 100	< 100
Industrial chemicals	> 10000	≪ 100
Nitrosamines	≦ 100	≪ 100
Natural toxicants	> 1000	< 50
Chemicals from packaging	> 1000	< 50
Pesticides	> 100	> 100

† Excluding structurally very similar substances.

toxicants and chemicals from packaging—have been studied quite intensively by food surveillance, but still there is considerable scope for further studies on their contamination of food. The most intensively studied group of organic chemical contaminants in food, pesticides, is also one of the smaller classes. These differences between the numbers of chemical contaminants that might be found in food and the total numbers in each category that have actually been detected in the diet illustrate the varying degrees of development in scientific study of the different contaminants. Table 1.3 does not include inorganic contaminants. There has been considerable work on several metals in food, notably lead, tin, cadmium, mercury, aluminium, copper and zinc—amongst these most effort has probably been on lead (Chapter 7). There has been much less research on other non-radioactive metals in food, and little work on non-metallic elements, except the metalloid arsenic—which is a classical contaminant—and elements such as iodine which are of interest to nutritionists.

1.4 CONTAMINATION ROUTES

There are many different ways in which food can become chemically contaminated: during its production, manufacture, processing, preparation, treatment, packing, packaging, transport and storage. Thus any scheme can only summarize the routes. Fig. 1.1 summarizes some of the routes of chemical contamination of plant-based foodstuffs, for example cereal grain products, as they move along the food chain. The complexity of routes, illustrated in Fig. 1.1, has profound consequences for the ways in which chemical contaminants in food are studied:

- It is essential that the exposure of consumers to chemical contaminants in food is quantified by carefully targeted surveillance.
- It can be difficult to trace problems of food chemical contamination to source—considerable resources can be required to identify the routes involved. If more than

one contamination route is implicated the fluxes of contaminants through the various pathways should be quantified so that the primary ones can be identified.

There are also implications for the control of chemical contamination of the food supply:

— Since there are several stages in food production, each with at least two possible sources of chemical contamination, one tier of control (at, say, the level of food processing) is unlikely to be fully effective. In practice control is often exercised at several points in the food chain.
— Because it can be difficult to identify all the sources of contamination, control is usually applied to the main identified source or sources.

Thus it is only possible to carry out meaningful surveillance and control on chemical contaminants in the diet if resources are effectively and pragmatically applied.

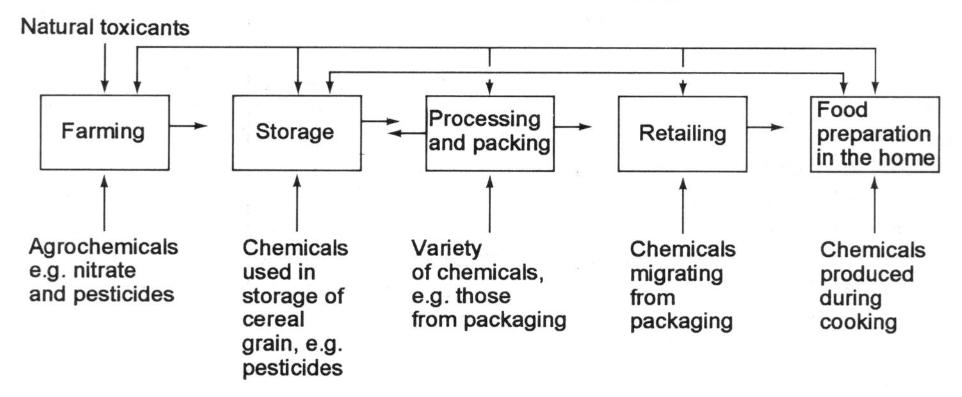

Fig. 1.1. Summary of some routes of chemical contamination of plant-based foods.

1.5 CONTINUING SCIENTIFIC WORK

Scientific research on chemical contaminants in food is still at an early stage. Not only is there a clear need to define the full range of chemical contaminants that is found in food (section 1.3) but there must be further progress in developing the methodology used. Analytical methods continue to improve, but there also needs to be greater emphasis on the sampling procedures used to obtain food and drink samples, if our estimates of consumer exposure are to be accurate.

Similarly there will inevitably be continuing extensive development of toxicology, but a concentration of effort on the following aspects of chemical contaminants would be very helpful: how is their toxicity changed when contaminants bind to macromolecules? and what are the net toxic effects of mixtures of different contaminants? (These and other areas that are in particular need of further scientific research are discussed in Chapter 11.) It is important to bear in mind when reading this book that there are several areas of work where too little is known about chemical contaminants—our current picture about many groups of chemical contaminants in food may be quite extensive but it is probably far from complete.

2

Veterinary drug residues

S. N. Dixon, D. R. Tennant and J. F. Kay, Ministry of Agriculture, Fisheries and Food, Rooms 215, 235 and 217, respectively,Ergon House, c/o Nobel House, 17 Smith Square, London SW1P 3JR, UK.

2.1 INTRODUCTION

Most people are aware of the considerable advances that have been made in the past forty years in providing medicines for the effective treatment of human diseases. Fewer are acquainted with a similar evolution of pharmaceutical products for use in animal husbandry for human food production. As with human medicines, veterinary drugs must undergo extensive development and testing to ensure that they are both effective and safe to use. A further consideration, which must be borne in mind when developing veterinary drugs, is the possibility that residues may remain in animal products which enter the human food chain and the potential consequences of the presence of such residues.

This chapter considers the ways in which veterinary drugs are used, the ways that such uses are controlled, the surveillance for residues and the possible effects on human health of veterinary drug residues in food.

2.1.1 Veterinary drugs on the farm

In 1991 in the UK there were approximately 12 million cattle, 43 million sheep and lambs, 7 million pigs, 127 million fowl (including chickens and laying hens) and 10 million turkeys, as well as many other farm animals and fish being raised for human food production. In the same year approximately £170m of veterinary pharmaceuticals were sold, mainly into agriculture. Veterinary drugs have important roles to play in reducing the disease and suffering of animals, controlling the hazards of disease transmission to man, increasing the productive capacity of animals and assisting in the efficient management of large numbers of animals on the farm.

2.1.2 Therapeutic agents

Therapeutic agents are used to control infections by pathogenic bacteria, fungi, or infestations by parasites. In such therapeutic uses, doses are determined to rid the animal of the disease-causing organism while causing no long-term health effects in the animal itself.

Therapeutic doses of veterinary drugs are usually administered to individual animals, often by injection, by a veterinary practitioner. In cases where large groups of animals are affected it is often more practicable to administer the drug orally in animal feed or drinking water.

2.1.3 Prophylactic agents

Prophylactic use of drugs as a precaution against infection is often employed when animals are known to be prone to infection. The drugs are given in order to prevent outbreaks of diseases in particular circumstances, for example at a certain time of year. Farmers are sometimes accused of using prophylactic drug treatment as a substitute for good husbandry. However, it is notable that the development of intensive animal production systems would not have been possible without the availability of prophylactic drugs. Products which can be administered in drinking water, feedstuffs or by dipping animals, to entire flocks or herds are the most attractive. However, this route of administration is the most difficult to control and has the highest potential for creating problems with drug residues in food, because it is difficult to control the amount of drug received by each animal and to separate animals under treatment from those due for slaughter.

2.1.4 Growth promoting agents

Growth promoters fall into two classes: antimicrobial and anabolic agents. Antimicrobials mixed with feed at sub-therapeutic concentrations suppress the activities of some of the natural bacteria living in the animal's intestinal tract. This results in improved feed utilization, and animals gain weight more rapidly. Anabolic growth promoters exert their effects via the animal's metabolism. Before anabolic growth promoters were banned in the United Kingdom they were usually administered as a pellet implanted in the animal's ear, the implant releasing the growth promoters gradually over a period of time. The possibility of food contamination was reduced by discarding the animal's ear at slaughter.

2.1.5 Herd and flock management

Hormonal drugs are used to control reproduction in many farm animals, by regulating fertility, in breeding programmes and to control parturition. Animals would not normally be slaughtered after such treatment. However, in the absence of a withdrawal period (the period between administering the drug to the animal, and taking milk or eggs or slaughtering the animal for human consumption) it is possible for dairy cattle prior to calving to produce milk which may have contained veterinary drug residues. Withdrawal periods are therefore defined when drugs are licensed. Tranquillizers may be given to animals to curb excitement or to control aggressive behaviour. Such drugs could be misused to control stress on transportation to slaughter, and residues of drugs used for this purpose would remain at high concentrations in edible tissues. It is essential to observe the instructions for using these and other drugs to avoid this sort of problem.

A number of drugs can be employed for more than one purpose. Some examples of these are mentioned in section 2.3. However, the use of therapeutic antimicrobials as growth promoters is restricted to avoid the possibility of pathogenic organisms developing resistance to these drugs. The antimicrobial agent would then be of no value in controlling outbreaks of disease caused by the resistant micro-organism.

2.2 CONTROL OF VETERINARY PRODUCTS IN THE UK

2.2.1 The Veterinary Products Committee

In the United Kingdom, under the *Medicines Act 1968*, the licensing authority for veterinary drugs is the Ministry of Agriculture, Fisheries and Food (MAFF). MAFF is advised by an independent body of experts, the Veterinary Products Committee (VPC). The VPC includes expertise from a wide variety of disciplines including veterinarians, the medical profession, toxicologists, microbiologists and agricultural specialists. It can also seek advice from other expert bodies from within MAFF, the Department of Health and elsewhere.

Pharmaceutical companies wishing to register new products must submit a dossier of data, relating to trials and investigations on the product, to MAFF. When the Ministry is satisfied that the data are complete and the research has been properly conducted they are passed to the VPC. The VPC then assesses the safety, quality and efficacy of the product. If there are any substantial doubts, the licence is refused and the company advised to carry out further investigations.

One of the principal aspects that the VPC considers is the safety of residues to the consumer. This is achieved by consideration of the toxicological data and setting a no effect level (NEL ≡ No Observed Effect Level) for that substance in an experimental animal. The NEL together with an additional safety factor, to allow for any interspecies and intraspecies variability, are used to calculate the acceptable daily intake (ADI) (see Glossary). For an adult human the ADI is calculated from the NEL by using the following formula for a 60 kg person:

$$\text{ADI} = \frac{\text{NEL} \times 60}{n}$$

where n = safety factor.

A safety factor of 100 is usually used for an NEL based on the combined toxic effects of a drug residue. Where teratogenic effects (birth defects) are involved a safety factor of 1000 is employed. When licensing veterinary drugs the ADI is compared with the amount of the substance that would occur in the daily diet of a consumer if the products he or she were eating had come from an animal treated with the drug. Only if the daily intake of an *extreme* consumer of the substance is less than the ADI would a product licence be granted. If the intake was greater than the ADI or the VPC was dissatisfied for some other reason about residues aspects, the pharmaceutical company would be advised to increase the recommended withdrawal period (section 2.1.5) so that residues would be reduced or the product containing the drug could be refused a licence.

It is up to the applicant for a licence to demonstrate that residues levels have declined, through the processes of metabolism and excretion, to a safe concentration within the proposed withdrawal period. The withdrawal period must then be stated on the product data sheet which accompanies the product. Veterinarians and farmers must observe these withdrawal periods if safe concentrations of residues are not to be exceeded. Many withdrawal periods are recommendations and are not currently binding in law, although the recent introduction of legislation controlling maximum residue limits (MRLs) for certain veterinary drugs necessitate the strict observance of product data sheets and hence withdrawal periods for those compounds included in the legislation. As a further safeguard it is important to carry out surveillance for residues in food (section 2.4).

2.2.2 Determining withdrawal periods

When a veterinary drug is administered the concentration in the animal's tissues will tend to rise with time until a plateau is reached. At this point the rate at which the drug is absorbed will be equal to the combined effects of metabolism and excretion in eliminating the drug from the animal. The concentrations of the drug in different tissues will not be equal as different types of tissue (kidney, muscle, liver, etc.) will have different affinities for the drug. Drugs are sometimes designed to 'target' particular tissues if they are the sites of infection. Concentrations of residues are frequently higher in liver, because it is the principal organ of metabolism, and in kidney, because of its role in excretion. When administration of the drug ceases, concentrations in tissues gradually decline as the drug is eliminated.

In theory this decline will be near to exponential (Fig. 2.1). In practice considerable variations from this occur because tissues with a high affinity for the particular drug will retain residues longer, and because individual animals have different rates of metabolism and excretion.

When pharmaceutical companies carry out experiments to determine withdrawal periods they normally use drugs which have a radioactive isotope incorporated into their structures. The presence of the radioactive isotope is harmless to the animal but can be detected by radiotracer techniques. This approach permits the determination of the

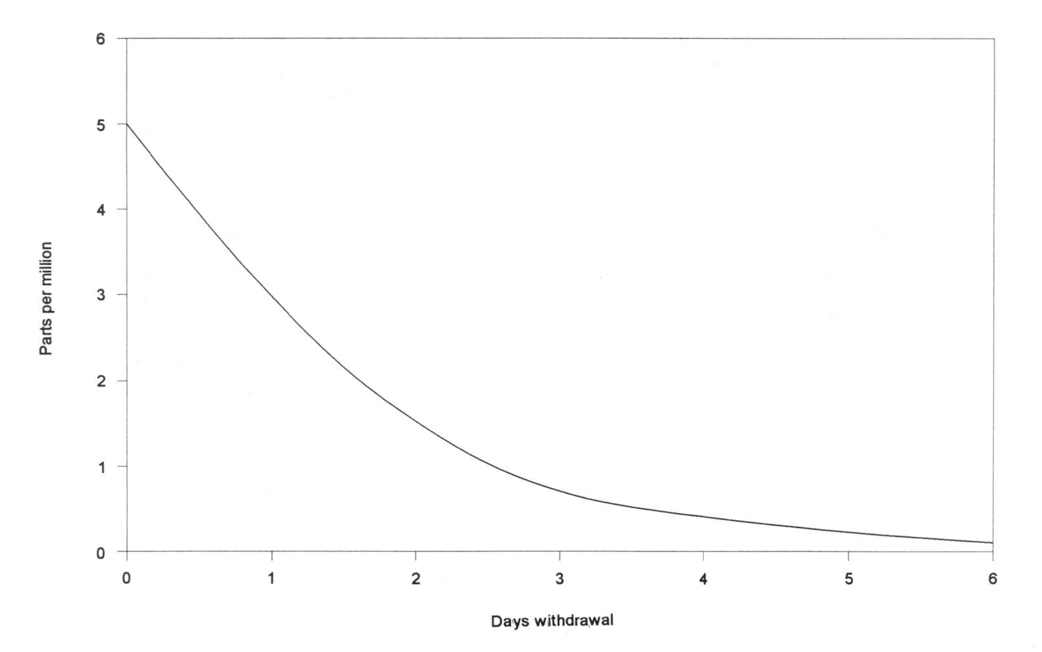

Fig. 2.1. Simple withdrawal curve.

biological half-life of the drug and the total residue concentrations in different tissues in different animals.

The withdrawal period is set at the time taken for residues to fall to a concentration below the MRL in each tissue (where this has been set), and allows a reasonable margin of error to allow for individual variations. For drugs where no safe concentration can be defined, for example in the case of some carcinogens, the limit of determination of the best analytical method available might be used.

2.2.3 The influence of EC controls

United Kingdom legislation on veterinary drug residues is very closely related to regulations and directives that are agreed for the whole European Community (EC):

— The introduction into UK legislation of maximum residue limits (MRLs) for several veterinary drugs (SI 1991/2843) was in anticipation of similar legislation to be introduced throughout the EC (Council Regulation No. 2377/90). Under this legislation, it is an offence to administer an unlicensed medicine to a food-producing animal or to sell or supply for slaughter an animal containing residues in excess of the prescribed MRL. This also applies to animals submitted for slaughter within the specified withdrawal period for a particular veterinary drug and also to meat and meat products.

— EC Council Directive 86/469/EEC concerns the examination of animals and fresh meat for the presence of residues of veterinary medicines, contaminants and other substances. The Directive was adopted in order to standardize monitoring for such residues in EC Member States. Regulations were passed to implement the Directive in the UK in July 1988. The Directive acts in two ways. Firstly it is an offence for the farmer to fail to keep adequate records of drugs administered to animals in their charge. The records must include the name of the drug, its dosage and the dates on which it was applied. The second way in which the Directive acts is to stipulate sampling schemes for all cattle, pigs, sheep, goats and horses arriving at slaughterhouses and destined for human consumption. The rates of sampling are largely based on the throughputs of animals through slaughterhouses to the market. The rates of sampling in subsequent six-month periods will depend upon the results of sampling in the initial stage. The levels of monitoring to be applied are listed in Table 2.1. In the UK, statutory surveillance is undertaken for the Veterinary Medicines Directorate by the Central Veterinary Laboratory and the Department of Agriculture for Northern Ireland's Science Service laboratories under the National Surveillance Scheme (NSS). These analyses are carried out in fulfilment of the requirements of EC Directives 85/358/EEC and 86/469/EEC. It is an offence in UK law under the *Medicines (Hormone Growth Promoters) (Prohibition of Use) Regulations 1986* (S.I. [1986] No. 1876) to use any of the substances in Group AI (Table 2.1). Any residues of these substances that are detected will be traced back to the farm of origin and, if possible, prosecutions brought.

— Similar EC Directives are presently being drafted to extend monitoring to include products imported from non-EC countries and to cover poultry, farmed fish and game. When these Directives have been adopted monitoring for residues of veterinary drugs will be comprehensive.

Table 2.1. Summary of sampling requirements of Directive 86/469/EEC (numbers are of animals to be sampled in the given period)

Residue groups	Sampling stage	Animal groups		
		Cattle (less than 2 years)	Culled cows	Pigs, sheep & goats & solipeds
Group AI (a) Stilbenes, stilbene derivatives, their salts, and esters (b) Thyrostatic substances (c) Other substances with oestrogenic, androgenic or gestagenic action, with the exception of substances in Group II	Initial	At least 0.15%, of which at least 0.10% are slaughtered and the rest inspected at the farm	700	700
	Intensive	At least 0.25%, of which at least 0.10% are inspected at the farm	0.25% of slaughtered cows	Double the routine inspection. At least 0.1% of animals slaughtered
	Routine	300†	300†	300†
Group AII‡ Substances authorized in accordance with Article 4 of 81/602/EEC and Article 2 of 85/649/EEC	Initial	—	700§	—
	Intensive	—	0.25% of slaughtered cows	—

Table 2.1 (cont'd)

Residue groups	Sampling stage	Animal groups		
		Cattle (less than 2 years)	Culled Cows	Pigs, sheep & goats & solipeds
Group AIII (a) Antibiotics, sulphonamides and similar antimicrobial substances	—	←	0.10% of animals slaughtered	→
(b) Chloramphenicol	Initial	←	At least 0.01% of slaughtered animals up to a maximum of 300 samples of each species	→
	Routine	←	300[†]	→

Table 2.1 (cont'd)

Residue groups	Sampling stage	Animal groups		
		Cattle (less than 2 years)	Culled Cows	Pigs, sheep & goats & solipeds
Group BI: Other medicines				
(a) Endo-and ectoparasitic substances				
(b) Tranquillizers and beta-blockers				
(c) Other veterinary medicines				
Group BII: Other residues		←	700 (all animal groups, groups BI and BII)	→
(a) Contaminants present in feedingstuffs				
(b) Contaminants present in environment				
(c) Other substances				

† These samples to be taken in such a way as to give at least a 95% assurance that there will be residues in less than 1% of the animals if no positive findings are made.
‡ Substances with oestrogenic, androgenic or gestagenic action, for therapeutic use, synchronization of oestrus etc.
§ These samples to be taken in such a way as to give at least a 99.9% assurance that there will be residues in less than 1% of the animals if no positive findings are made.

— To standardize conditions further in the EC and to assist in the implementation of Directives, the European Commission is defining laboratory methods to confirm the presence of residues of veterinary drugs, and guidelines for screening methods (*Veterinary drug residues—residues in food-producing animals and their products: Reference materials and methods*. ISBN 92-826-4095-7). The requirements of these methods are extremely rigorous and should ensure that effective prosecutions can be brought under the Regulations implementing the Directive.

2.2.3.1 The EC Committee for Veterinary Medicinal Products' Working Group on the Safety of Residues of Veterinary Medicines

The EC Committee for Veterinary Medicinal Products (CVMP) was established to harmonize licensing procedures for veterinary products throughout the European Community. The CVMP recognized that residues of veterinary products could lead to barriers to trade in animal products if different standards were adopted in EC Member States. A working group on the Safety of Residues of Veterinary Medicines was established in order to agree common safety standards and prevent barriers to trade in such products within the EC. This Working Group is charged with recommending to the CVMP tolerances for veterinary drug residues in food. On the advice of the Working Group, the CVMP has set tolerances of 0.01 mg/kg for residues of chloramphenicol and 0.1 mg/kg for sulphadimidine (section 2.3.1). A wide range of veterinary products is currently under consideration by the CVMP.

2.2.4 Other international organizations

2.2.4.1 Codex Committee on Residues of Veterinary Drugs in Food

A large proportion of many nations' animal produce is destined for international trade. In order for consignments to be acceptable to importing countries, common safety standards must be adopted between exporters and importers. In the past the United Nations Food and Agriculture Organization (FAO) and World Health Organization (WHO) have collaborated to agree common standards for food additives, pesticides residues and some food contaminants. A Joint FAO/WHO Expert Consultation on Residues of Veterinary Drugs in Foods met in Rome in 1984 and on their advice the UN Codex Alimentarius Commission established the Codex Committee on Residues of Veterinary Drugs in Food (CCRVDF). The CCRVDF comprises delegates from UN Member Nations. The Committee takes expert advice from the Joint FAO/WHO Expert Committee on Food Additives (JECFA) on potential public health hazards and barriers to international trade arising from veterinary drug residues in food. The JECFA had previously considered the use and assessed the safety of certain veterinary drugs and their residues. The Committee had also made recommendations on the use of several antibiotics, and undertaken evaluations of residues of chloramphenicol and some natural and synthetic hormonal growth promoters. The CCRVDF works with JECFA to provide recommendations on the use of maximum residue limits and acceptable daily intakes of veterinary drugs. For example, when the CCRVDF met for the first time in Washington in 1986 it identified some priority substances. In response, JECFA met in 1987 to re-evaluate trenbolone acetate and zeranol, and evaluate chloramphenicol, oestradiol, progesterone and testosterone (section 2.3).

After assessing patterns of use, metabolism and pharmacokinetics, toxicological data, residue depletion (under field conditions) and analytical criteria for each of these compounds, JECFA proposed acceptable daily intakes (ADIs) for some of the compounds. Their recommendations are listed in Table 2.2. JECFA made no recommendations regarding ADIs or acceptable residue concentrations for chloramphenicol because it was unable to establish a no effect level for the toxic effect of this substance. It was recommended that the use of chloramphenicol should be prohibited, particularly in laying birds and lactating farm animals. The Committee considered that it was unnecessary to set an ADI for natural hormones as they are produced endogenously in human beings and show great variations in concentration according to human age, sex and physiological status. It was concluded that residues arising from the use of these compounds, in accordance with good animal husbandry practice, were unlikely to pose a hazard to human health. Although their administration by subcutaneous implant at the base of the animal's ear was considered, the potential hazard posed by the consumption of injection sites, following illegal use of these substances, was not reviewed.

Temporary acceptance of the ADI and acceptable residue concentrations for trenbolone acetate were recommended by the CCRVDF pending the submission of further data. The maximum permissible concentration of zeranol residues in meat was estimated to be 0.070 mg/kg of edible tissue. However, the Committee noted that concentrations in animals treated in accordance with good husbandry practice would be lower than this. Accordingly, lower acceptable residue concentrations were recommended, the limit for

Table 2.2. JECFA Recommended MRLs for certain veterinary agents

Substance	Acceptable daily intakes for human beings	Acceptable residue concentrations
Chloramphenicol	Not allocated	Not allocated
Oestradiol-17β	Unnecessary	Unnecessary
Progesterone	Unnecessary	Unnecessary
Testosterone	Unnecessary	Unnecessary
Trenbolone acetate	0 to 0.01 μg/kg of bodyweight†	1.4 μg/kg (bovine tissue) for β-trenbolone†‡ 14 μg/kg (bovine liver and kidney) for β/α-trenbolone†
Zeranol	0 to 0.5 μg/kg of bodyweight	10 μg/kg (bovine liver)§ 2 μg/kg (bovine muscle)¶

† Temporary acceptance.
‡ Based on 500 g of meat consumed per day by a 60 kg person.
§ Based on concentration consistent with good animal husbandry practice.
¶ Based on lowest concentration consistent with the practical analytical methods available for routine residue analysis.

bovine muscle being the lowest concentration consistent with the practical analytical methods available for routine residue analysis.

At the second meeting of the CCRVDF held in Washington in 1987 the Committee began the process of developing maximum residue limits (MRLs). A draft definition of the MRL was adopted as follows:

> '**MAXIMUM RESIDUE LIMIT (MRL)**—is the maximum concentration of residue resulting from the use of a veterinary drug that is recommended by the Codex Alimentarius Commission to be legally permitted or recognised as acceptable in or on a food.'

The MRL is based on the type and amount of residue considered to be without any direct or indirect toxicological hazard for human health. It is established on the basis of an ADI or, where this is not possible because of insufficient scientific knowledge, on the basis of a temporary ADI which uses an additional safety factor. It takes into account factors such as resistance promotion, allergenic potential and other undesirable side effects, which may have either a direct or indirect effect on human health. The MRL may be reduced to accommodate residues that originate in food of plant origin and/or the environment.

A definition of good practice in the use of veterinary drugs was also drafted:

> '**GOOD PRACTICE IN THE USE OF VETERINARY DRUGS (GPVD)** is the officially recommended or authorised usage approved by national authorities, of veterinary drugs under practical conditions in a manner that leaves toxicologically acceptable residues of the smallest amounts practicable.'

The CCRVDF adopted the recommendations of JECFA (Table 2.2) as draft MRLs. These draft MRLs have been reviewed by individual member nations.

2.3 CHEMICAL SUBSTANCES COMMONLY USED AS VETERINARY DRUGS

2.3.1 Antimicrobial agents

Antimicrobial agents are chemicals that selectively inhibit the growth of pathogenic micro-organisms, particularly bacteria. These drugs may be used therapeutically to treat disease, or prophylactically to prevent disease or to promote growth. Growth-promoting antimicrobials are considered by the VPC to be safe for inclusion in animal feeds at low concentrations, as the quantities used are considerably lower than are required therapeutically; as a result there is proportionately less concern over their residues. There is more interest amongst specialists in residues of those drugs used therapeutically in animals, which are also used in human medicines. Interest has focused particularly on chloramphenicol, the sulphonamides, the tetracyclines and the nitrofurans.

Chloramphenicol (Fig. 2.2) is particularly valuable in the treatment of systemic salmonellosis, particularly in calves, and respiratory infections in calves and pigs. However, in the UK, veterinary practitioners are advised to prescribe chloramphenicol only where no effective alternative therapeutic agent is available. This is because there is a particular toxicological concern about any residues that might arise from its use.

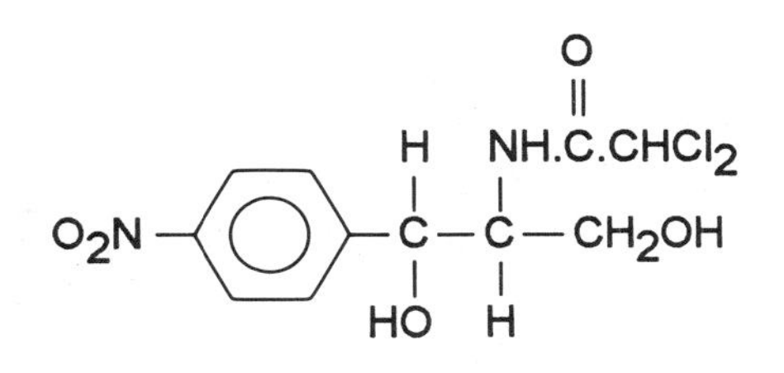

Fig. 2.2. Chemical structure of chloramphenicol.

The *sulphonamide* drugs (Fig. 2.3) are a family of substances derived from sulphanilamide and developed for the treatment of systemic bacterial diseases in human medicine; there are several thousand individual compounds. Although their use in human medicine has been gradually replaced by more modern antibiotics, they are still widely used therapeutically in animals because of their low cost, convenient (oral) administration and effectiveness. The use of sulphadimidine in pig farming is particularly widespread for the control of pneumonia resulting from *Haemophilus* infections, a disease which can cause severe losses when pigs are kept in enclosed conditions during winter months.

Tetracyclines (Fig. 2.4) are another large family of related compounds. They were developed after the sulphonamides and are widely used therapeutically in humans, animals and fish. They are also employed at sub-therapeutic doses as growth-promoting feed additives. There may be interest in residues from their use in fish farming, as relatively large doses may be required to control the spread of infection among salmon and trout kept in close confinement on fish farms.

Nitrofurans (Fig. 2.5) are a family of antibacterial agents used in poultry production. A number of nitrofurans have gained commercial success in human and animal medicine because of their broad spectrum activity against bacteria and since they tend not to induce drug-resistant bacterial strains. They are rapidly metabolized—hence it is residues of these metabolites that are of interest, not residues of the parent nitrofurans. However, their safety is in question and a number of expert committees are currently reviewing the toxicity data on these compounds.

2.3.2 Anabolic agents

Those anabolic agents used in veterinary medicine are naturally occurring steroids (Fig. 2.6). They are no longer permitted for use as growth promoters in the UK. The administration of hormone growth promoters was prohibited in the United Kingdom from 1 December 1986 under the *Medicines (Hormone Growth Promoters) (Prohibition of Use) Regulations 1986* (S.I. [1986] No. 1876). Although this type of growth promoter has now been banned throughout the European Community because of fears about health effects from residues and the need to standardize conditions of trade in animal products, they are still licensed for use in some other countries. The use of one particular group of synthetic anabolic agents, the stilbene compounds (Fig. 2.7), was prohibited in 1982 because of fears about possible adverse health effects from residues.

The steroid hormones are synthetic products which are either identical to endogenous (natural) male and female sex hormones (Fig. 2.6), or have similar structures (Fig. 2.7).

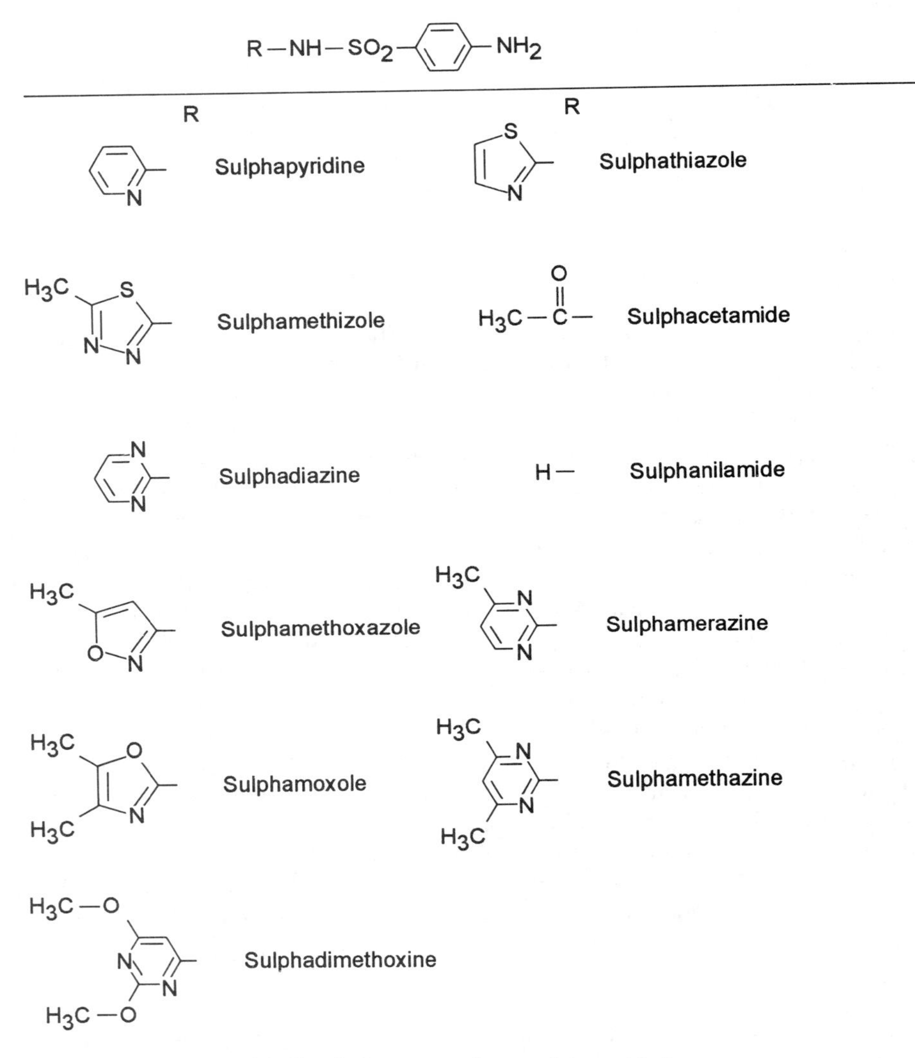

Fig. 2.3. Chemical structures of some sulphonamide drugs.

They have an important role in growing animals in the development of muscle tissues. They were administered as an implant to the base of the animal's ear. Their effect on the animal's metabolism results in a more efficient use of feed and leads to more rapid growth. At slaughter the ear was discarded to prevent contamination of food with the drug left in the implant. Both natural and synthetic hormonal growth promoters were once in widespread use in beef production in the UK to counter the effects of castrating bulls to make them more docile. The castrated male (steer) is deficient in endogenous hormones and the administered hormones were used to stimulate normal growth without

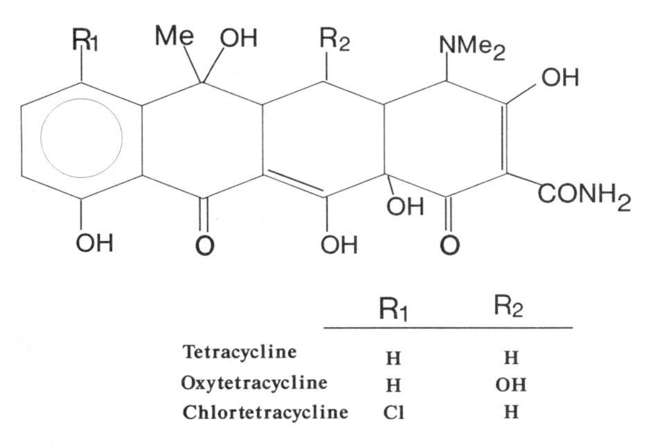

	R_1	R_2
Tetracycline	H	H
Oxytetracycline	H	OH
Chlortetracycline	Cl	H

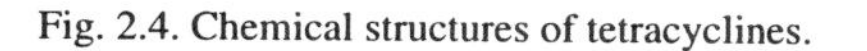

Fig. 2.4. Chemical structures of tetracyclines.

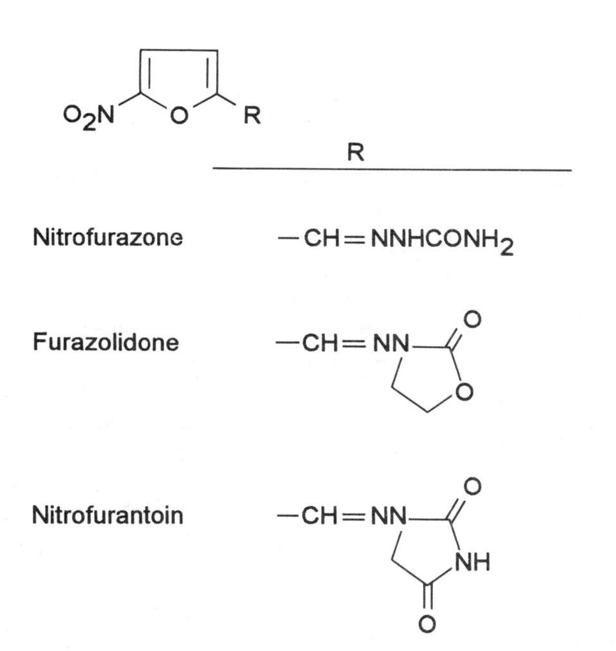

Fig. 2.5. Chemical structures of nitrofurans.

inducing the steer's aggressive tendencies. Female cattle (heifers) were also given hormones to increase their growth rate towards that of intact male animals.

In healthy, non-castrated animals sex hormone concentrations can vary widely. Studies have been undertaken to compare concentrations of naturally occurring hormones with those resulting from implants. Although the data (Table 2.3, p. 24) are incomplete, there is some evidence that residue concentrations of progesterone in fat from treated male calves may be higher than those found naturally. However, it is generally accepted

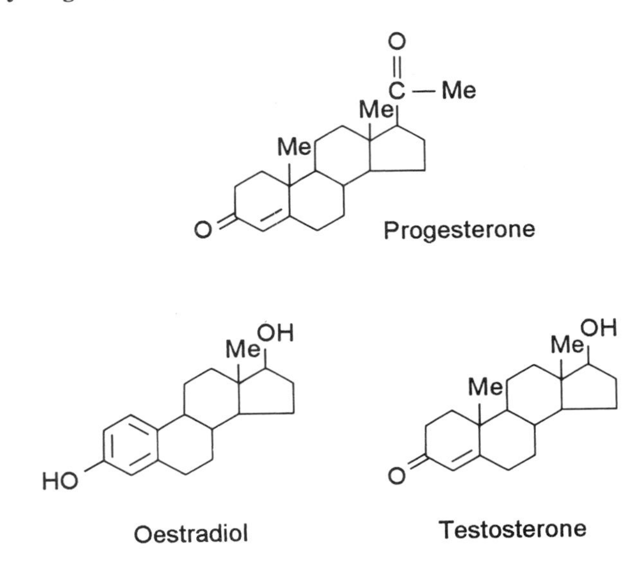

Fig. 2.6. Chemical structures of some natural hormones.

that concentrations of residues of hormones in the tissues of treated castrated animals are lower than those which occur naturally, provided that the hormones had been administered as an implant and the appropriate withdrawal period observed. Nevertheless the concentrations of natural hormones in treated animals are most usually within the range of concentrations found for the same hormones in untreated animals. The apparently normal concentrations of endogenous hormones in the tissues of treated animals makes the detection of illegal use particularly difficult.

The synthetic hormones, like the natural substances, can be administered in preparations either singly or in combination, for example an androgen (male hormone) and an oestrogen (female hormone), in order to obtain a combination of adult male and female growth characteristics. The stilbenes and zeranol (Fig. 2.7) are oestrogenic whilst trenbolone derivatives are androgenic. The total androgen and oestrogen concentrations in muscle from animals treated with combinations of these substances have been estimated to be lower than the combined levels of these hormones found in bulls or pregnant cows. However, detection of illegal use is simpler because the compounds do not occur naturally and thus the presence of any residue is evidence of illegal use.

Progestogens, such as progesterone, and oestrogens have also been used in animal husbandry for herd or flock management to synchronize animals coming into 'season', so that breeding programmes can be planned, and to induce milk production.

2.3.3 Anthelmintic agents

Anthelmintics are used to prevent and treat parasitic diseases in food-producing animals, particularly those ailments caused by gastrointestinal and lung worms, and liver flukes. In the past the most important compounds have been levamisole and the benzimidazoles

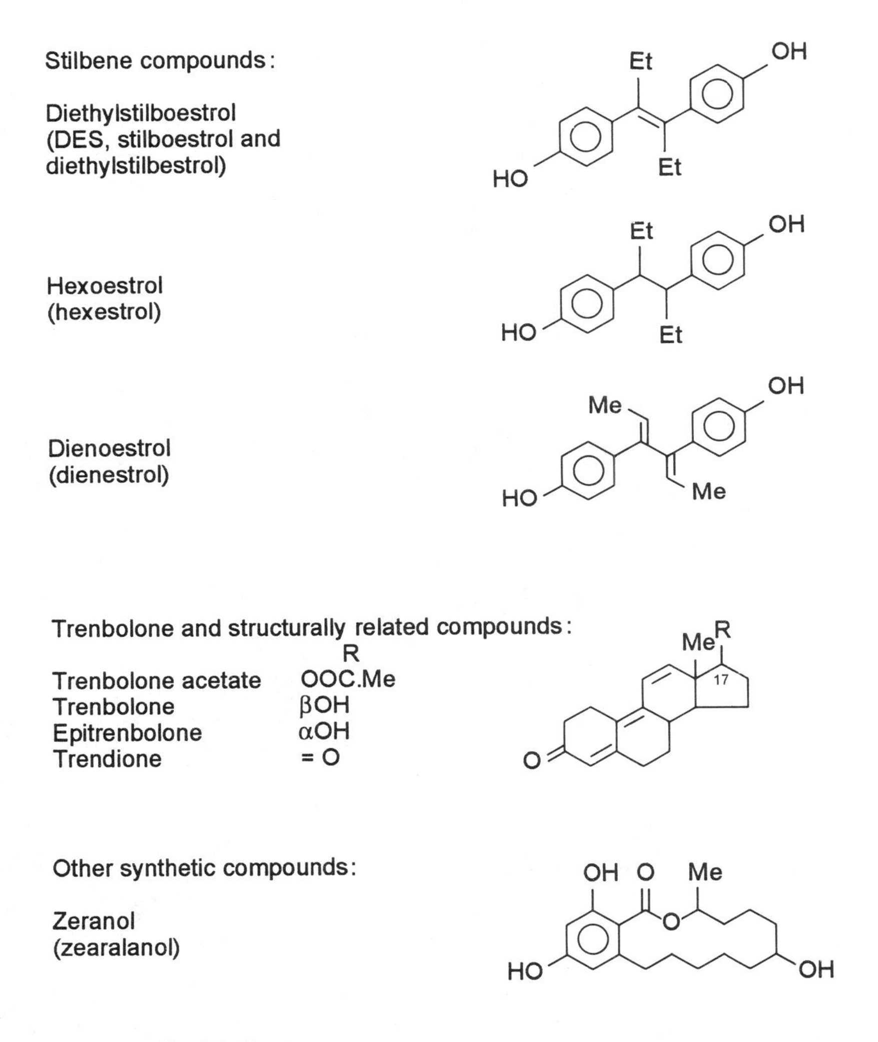

Fig. 2.7. Chemical structures of some synthetic hormones.

(Fig. 2.8, p. 25). These substances have been largely replaced by ivermectin (Fig. 2.9, p. 26) which has gained a large share of the market in many countries. Ivermectin has a broad spectrum of activity against parasites and is used in cattle, pigs and sheep. Anthelmintics are frequently used prophylactically when animals are turned out on pasture which may harbour dormant phases of parasitic organisms from the previous season.

2.3.4 Coccidiostats

The ionophores (Fig. 2.10, p. 26) are the largest group of compounds used for the control of coccidiosis. These compounds are widely used to control coccidial gut parasites in young

Table 2.3. Average concentrations (pg/g) of natural hormones in tissues of treated and untreated animals

Hormone	Animal	Muscle	Liver	Fat
Testosterone	Bull	535	749	10950
	Heifer	92	193	250
	Veal calf	16	39	685
	Veal calf (treated†)	70	47	340
	Steer	101	NM	NM
Progesterone	Pregnant cow	NM	NM	360200
	Heifer	NM	NM	16700
	Veal calf	NM	NM	5800
	Veal calf (treated‡)	NM	NM	12500
Total oestrogen	Veal calf	2	600	NM
	Veal calf (treated§)	7	840	NM
	Steer	12500	27600	NM
	Steer (treated¶)	21800	NM	NM
	Heifer	13000	71000	NM

† 77 days after implantation of 20 mg oestradiol-17β and 200 mg testosterone.
‡ 70 days after implantation of 20 mg oestradiol-17β and 200 mg progesterone.
§ 60 days after implantation of 20 mg oestradiol benzoate and 200 mg testosterone proprionate.
¶ 66 days after implantation of 20 mg oestradiol benzoate and 200 mg testosterone proprionate or 200 mg progesterone.
NM: not measured.

young poultry. As the birds mature they develop resistance to the parasites. In addition these compounds may also be used as growth promoters in feed for beef cattle.

2.3.5 Tranquillizers and beta-adrenergic agonists

Tranquillizers and beta-adrenergic blockers may have been used illegally in some countries to control stress in animals being transported and while awaiting slaughter. Pigs are particularly sensitive to sudden changes in their environment and the metabolic consequences of stress can adversely affect meat quality. The tranquillizers most likely to have been used for this purpose are azaperone, azaperol and propriopromazine. The beta-agonist compound clenbuterol also has a tranquillizing effect. This group of compounds acts by impeding the uptake of adrenal hormones by nerve cells and stimulation of the cardiovascular system. When treatment is prolonged (three to four months), they also induce a redistribution of fat to muscle tissue. There is evidence from Europe that this effect has been illegally utilized as a substitute for muscle generation via the now banned hormonal growth promoters. However, the quality of the meat produced in this way is poor. In the UK recently introduced regulations permit a concentration of 0.5 μg/kg in edible tissues, but only if a veterinary prescription is available to certify legitimate use.

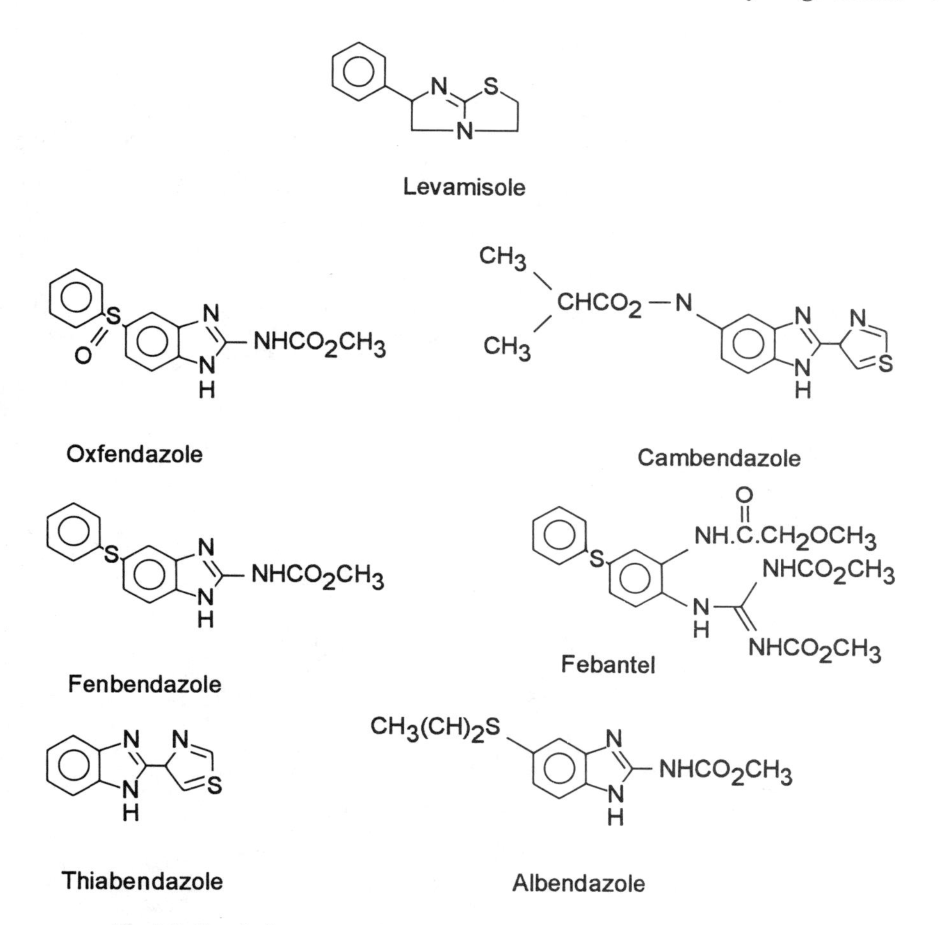

Fig. 2.8. Chemical structures of levamisole and some benzimidazole anthelmintics.

2.3.6 Non-hormonal growth promoters

The quinoxaline di-*N*-oxides, carbadox and olaquindox (Fig. 2.11) are used as growth promoters in pigs. As both of these compounds are rapidly metabolized surveillance for residues must include their principal metabolites, quinoxaline carboxylic acid (QCA) and methyl-QCA.

2.4 SURVEILLANCE FOR VETERINARY DRUG RESIDUES

2.4.1 Introduction

In a climate of some concern among consumers and food safety specialists about possible harmful effects of veterinary drug residues, there is a rapid growth in organizations, within developed nations, dedicated to surveillance for these residues. Veterinary drug residue surveillance is a relatively new science and much attention is currently being

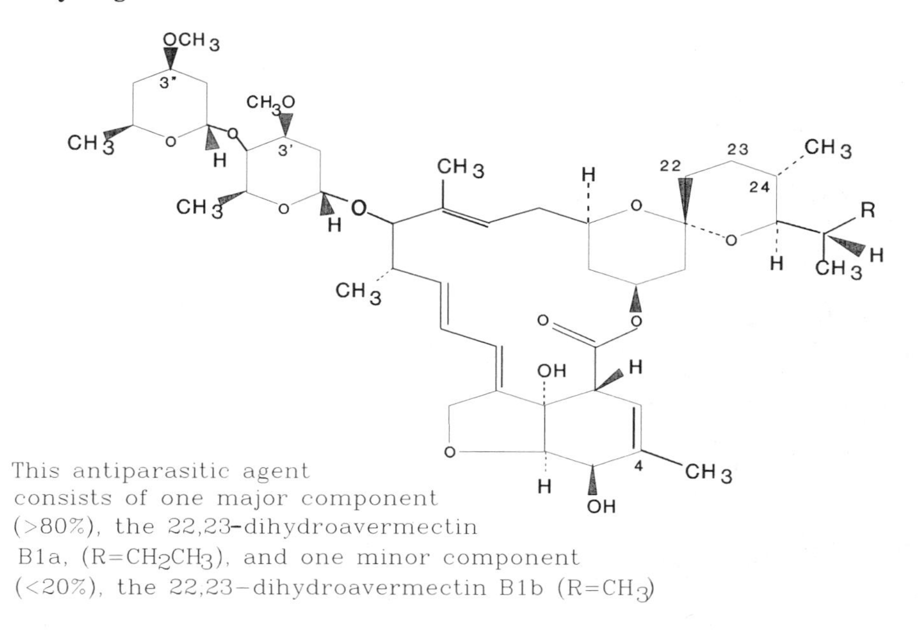

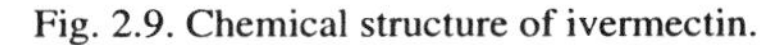

Fig. 2.9. Chemical structure of ivermectin.

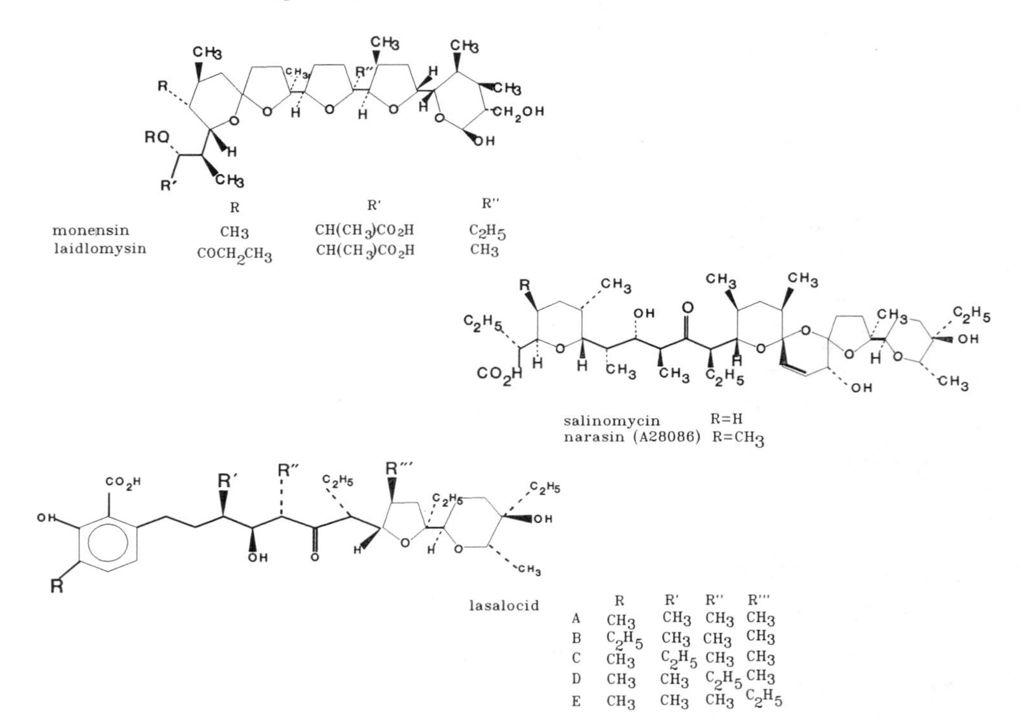

Fig. 2.10. Chemical structures of some ionophore coccidiostats.

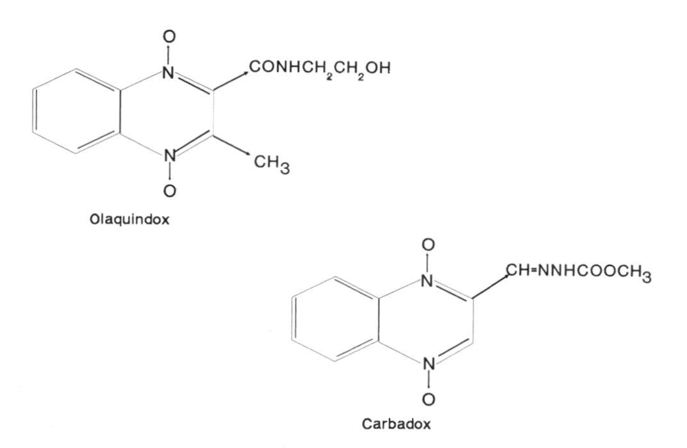

Fig. 2.11. Chemical structures of carbadox and olaquindox.

devoted to developing analytical methods and, where necessary, legislation, which will provide effective mechanisms for assessment and control.

2.4.2 National surveillance programmes

In the UK responsibility for advising ministers on the safety of the national food supply rests with the Steering Group on Chemical Aspects of Food Surveillance (SGCAFS) (Chapter 9, section 9.7). The Steering Group has delegated responsibility for detailed work on veterinary drug residues to its Working Party on Veterinary Residues in Animal Products (WPVR). The WPVR comprises Government scientists and administrators, independent experts and representatives from industry. The Working Party co-ordinates practical work to determine the incidences and concentrations of veterinary drugs in food, and advises the SGCAFS. Reports prepared for the SGCAFS are published periodically as food surveillance papers, the most recent report being Food Surveillance Paper No. 33 (published by HMSO, London, 1992). Surveillance co-ordinated by the WPVR—which is complementary to statutory surveillance conducted by the State Veterinary Service (SVS) in compliance with EC Directives—ensures that consumers are not exposed to concentrations of residues which may pose a potential hazard to health, and protects consumers against undue exposure to such residues. For example the data from the National Surveillance Scheme (NSS)—through which EC Directive 86/469/EEC (section 2.2.3) is enforced in the UK—are reported to the WPVR.

In other nations similar bodies exist for the surveillance of veterinary drug residues in food. In the USA this role is undertaken by the Food Safety and Inspection Service of the US Department of Agriculture. The US Food and Drug Administration has jurisdiction over poultry and livestock up until slaughter and defines safe tolerance limits for residues.

2.4.3 Surveillance for drug residues in food

This work may be undertaken with more than one objective in mind:

— Testing for compliance with nationally or internationally set food safety standards.

— Checking on the effectiveness of licensing and other control procedures.
— Estimating the exposure of consumers to veterinary drug residues in their diet.

These objectives are not comprehensive but the approach taken in surveillance will depend largely upon the desired objective (Chapter 9 reviews the general approaches involved). In the UK the WPVR aims to produce surveillance data which can be used to estimate dietary intakes. This is achieved by adopting a targeted approach. Consultation with toxicologists and veterinary experts identifies drug/food combinations where residues are likely to occur and which might theoretically present a potential hazard to human health. In such cases 'pilot', short-term, surveys are undertaken to estimate the extent of the potential problem. If residues are detected then more extensive surveillance is undertaken to quantify the risk. The WPVR also commissions research to identify potential problem areas. Trials are commissioned to indicate the concentrations of residues which would occur in animals slaughtered with and without the withdrawal period being observed, after administering the drug at the recommended dose. The effects of cooking on residues concentrations are also considered, because while cooking may destroy some compounds it may also release others which have become bound to tissues (this is discussed in more detail in Chapter 11).

The number of samples that needs to be analysed to determine the incidence of residues depends upon the objective of the surveillance and the nature of the potential harmful effects. For veterinary residues surveillance it has been possible to define the numbers of samples that need to be analysed (but see Chapter 9, Table 9.3). Table 2.4 shows how the number of samples required to detect residues with a given degree of certainty increases as the incidence of positives declines. An incidence of 1% with a confidence limit of 95% is the minimum standard that should probably be required for compliance purposes and so 300 samples per year is recommended. When surveillance is undertaken for food safety purposes the standard required will vary. For a compound where a single acute exposure could prove harmful many more samples will be required to confirm the safety of the food supply than for a compound where a long-term chronic exposure is needed to produce harmful effects. In the latter case an estimate of

Table 2.4. Numbers of samples required to detect at least one positive sample at given probabilities

Incidence of positives (%)	Minimum numbers of samples required to detect one positive with a confidence limit of:			
	90%	95%	99%	99.9%
20	11	14	21	31
10	22	29	44	44
1	230	300	459	688
0.1	2302	2995	4603	6905
0.01	23025	29956	46050	69075

the average exposure is required. Even so, a survey of less than 100 samples is rarely considered adequate.

For food safety purposes the overriding aim is that food contamination should be reduced to the lowest practicable level, bearing in mind the potential costs and benefits involved. Since it is difficult to establish cause and effect relationships following long-term exposures at low concentrations, it may be necessary to base action on prudence rather than on proven harm to health. However, if this approach is to maintain the confidence of both consumers and producers of food, a rational evaluation of all relevant information is required so that the balance between the risks and benefits of veterinary drugs can be assessed. Whilst information on the incidence of potentially harmful drug residues is fundamental to this cost-benefit analysis, so too is the consumption of the commodities involved (particularly by extreme consumers), the potential loss of food production if the drug is controlled or prohibited, and the animal welfare implications of restricting an important animal medicine for which there may be no effective substitute.

2.5 ANALYTICAL METHODS EMPLOYED IN DRUG RESIDUES SURVEILLANCE

2.5.1 Introduction

Veterinary drug residues tend to be present in edible tissues at very low concentrations. It is therefore essential to confirm the presence of a detected residue by a different analytical technique from the one used to detect the residue. Although the majority of samples analysed may give negative results, when a positive result is obtained a rigorous analytical approach must be adopted to establish or confirm it. A two-tier analytical approach is adopted by the WPVR in its surveillance programmes in the UK. Screening is used to detect suspect positive samples which are then retested and quantified by a different and more specific technique.

2.5.2 Screening analyses

Ideally, screening techniques should be relatively inexpensive, rapid and permit a large number of samples to be analysed. For veterinary drug residues analysis the basic criteria that the screening method must meet are:

- A limit of determination below the MRL, or as low as possible where no MRL has been set.
- A low incidence of false negative results (Chapter 9, section 9.2).

It is also desirable that a screening test should give a low incidence of false positive results. False positives will occur if the test is sensitive to other, similar compounds—such as natural substances in the tissue—or metabolites. A high incidence of false positives will limit the cost-effectiveness of the screening method because unnecessary expensive confirmatory tests will need to be performed.

In practice, screening methods often give only qualitative (i.e. positive/negative) or semi-quantitative (i.e. high/medium/low/negative) results. They are sometimes sensitive to more than one compound in a related group, giving either separate indications for each residue or a total for the group.

2.5.3 Confirmatory analyses

Confirmatory techniques must conform to more stringent criteria than screening techniques and have:

— a low limit of detection;
— a high level of accuracy;
— a high level of precision; and
— a high level of specificity to the drug in question.

It is important that there is a low level of doubt about the identity of the compound being measured and a high level of certainty that the quantity determined is a true reflection of the amounts actually present. To these ends, confirmatory techniques usually employ complex separation procedures to isolate the compound of interest and require calibration procedures which involve adding known amounts of the compound to uncontaminated specimens of the material under analysis. As a final check on the validity of the method it is recommended that a standard reference material should be analysed alongside the material under test. Standard reference materials are tissue specimens to which the compound has been added; or they can be prepared by slaughtering animals treated with the drug and sampling the same tissues as those being analysed. They will have been analysed with fully validated methods in different laboratories, to produce a consensus estimate of the concentration of the drug. The value obtained from an analysis of these samples can be used to confirm the validity of results achieved in an individual laboratory. At the time of writing the number of available standard reference materials is very low. However, the European Commission Bureau of Community Reference Materials (BCR) is developing certified reference samples for use in veterinary drug residue analysis.

2.5.4 Selection of tissues for analysis

When selecting a method for determining veterinary drug residues some knowledge of the metabolism of the drug is required. All constituents of a residue are not equally toxic and some may have longer biological half-lives than others. The most toxic metabolite may be transitory and present only in small quantities. The analytical method selected must be able to detect the residue in whatever form is most prevalent and/or toxic. For example, carbadox (Fig. 2.11) is rapidly metabolized via desoxycarbadox to quinoxaline carboxylic acid (QCA). In the UK and USA surveillance for residues of carbadox is achieved by analysis for QCA.

Some drug residues become bound to the tissues of the animal. This may have the effect of deactivating the toxic potential of the residue. However, determining the *total* concentration of the parent compound and/or its major metabolites generally ensures the highest margin of safety.

A knowledge of drug metabolism will also give an indication of which tissues are most appropriate for analysing in surveillance. Tissues do not accumulate residues to the same degrees. The roles of the liver and kidney in metabolism and elimination mean that these tissues frequently have the highest residue concentrations and are thus often selected for analysis. However, bile has been used for monitoring residues of anabolic hormones for regulatory purposes, because the bile duct is the major excretory route for hormones and concentrations in bile tend to reflect concentrations of these drugs in the liver. Any detectable residue of synthetic hormones in bile is evidence of a breach of the prohibition of the use of such substances. The position with natural hormones is less clear since endogenously produced hormones will always be present and administered natural hormones tend to give residues concentrations within the normal range (section 2.3.2). There are indications that determining a range of hormones to give a 'hormone profile' may provide an indication of natural hormone administration. This is because the administration of hormones may have an effect on the animal's secretion of endogenous hormones by influencing natural feedback mechanisms.

2.5.5 Analytical methods used to detect residues of antimicrobial agents

Microbial inhibition tests are widely employed in screening for antimicrobial residues. Such tests detect the antimicrobial activity of residues present in a tissue sample. The Four-Plate Test (FPT), also known as the Frontier Post Test, is typical. In the FPT, discs of tissue are placed on four agar plates inoculated with micro-organisms. The plates are incubated under different conditions to allow inhibition of growth by a variety of antimicrobial drugs. A positive result is indicated by complete inhibition of growth on the surface of the medium in a zone not less than 2-mm wide around the tissue disc. If a positive is confirmed then the identity of the residue must be established by a second technique. The FPT is only relevant for residues where the toxic risk to the consumer is related to the antimicrobial activity. The presence of toxic metabolites or bound residues which have no antimicrobial activity will not be detected. This could present a particular problem with residues such as the penicillin metabolite penicilloyl-protein complex which can cause allergic reactions but lacks any antibacterial activity.

The FPT also generates false positive results (section 2.5.2) which are thought to be due to non-specific inhibitory effects of natural constituents of the tissue. A further deficiency of the FPT is that it is not sufficiently sensitive towards some of the antimicrobial agents that are available for use in food-producing animals. Thus the FPT has some value in indicating the degree of compliance with recommended withdrawal periods for antimicrobial agents but it has very limited value in assessing health risks from residues of such agents.

Because of the deficiencies of the FPT, the WPVR has placed great emphasis on the development of quantitative chemical methods for the analysis of residues of antimicrobial residues. Chemical methods for drug residues usually require the homogenization of the tissue and then a laborious extraction of the compound to be determined. The constituents of the sample extract are separated by chromatographic techniques. Gas-liquid chromatography (GLC) or high-performance liquid chromatography (HPLC) are most frequently employed. Detection in GLC is usually by flame ionization (FID), nitrogen/phosphorous (NPD), or electron capture (ECD) detectors. Mass spectrometry (MS) can also be employed to give a unique fingerprint of the chemical composition of the residue and is becoming more widespread in use. High-performance liquid chromatography detection systems are usually based on ultraviolet and fluorescence spectroscopy or electrochemical detection. Chromatographic analyses are often able to detect a range of chemically related drug residues. This is particularly the case with 'families' such as the sulphonamides and tetracyclines. Such multi-residue techniques have obvious cost zadvantages and are employed as screening tests wherever possible. Their use will undoubtedly become more widespread—this is already the case for the detection of pesticides residues (Chapter 8, section 8.3.3). Immunoassay techniques are finding increasing application in the determination of residues of antimicrobial agents. It is beyond the scope of this chapter to describe how such methods work (but see Chapter 9, Table 9.1). A recent development in drug residue surveillance has been to mount the immunoassay reagents on a card. Card-mounted tests are commercially available for chloramphenicol, sulphadimidine, and neomycin. These tests can be used to test for the presence of drug residues in liquid media, such as bile, serum or urine, on site and may be used for regulatory purposes at slaughterhouses.

The most time-consuming step in assays of tissues for drug residues, either by chemical analysis or immunoassay, is the preparation of a relatively pure extract that is suitable for analysis. This is because many natural constituents of tissues can interfere with these very sensitive analytical techniques. A recent approach to speeding up this process is by the use of immunoaffinity clean-up columns. Here an impure aqueous solution of the tissue extract is passed through a chromatographic column containing immobilized antibodies to the drug under analysis. The drug residues bind to the antibodies and all other materials are rinsed away. The drug residues are then desorbed using a different solvent and analysed by using one of the methods described above. The column can be reused many times and the process is suitable for automation. Immunoaffinity columns can be made to select a range of different or related compounds, which can in some cases be desorbed sequentially by altering the solvent.

Residues of hormonal growth promoters in tissues are usually present at very low concentrations and many chromatographic methods are too insensitive. Although thin-layer chromatography can be used to detect diethylstilboestrol and trenbolone in fluids, radioimmunoassay (RIA) is the most widely used method. For other drugs, where the concentration of residues is usually higher, chromatographic techniques are more widely used. There is increasing interest in the application of enzyme-linked immunoabsorbent assay (ELISA) and immunoaffinity columns in the detection of anabolic agents, anthelmintic drugs and other medicines. However, considerable resources must be invested in developing such techniques and it is likely to be some time before they become widely available for drug residues analysis.

2.6 RESULTS OF SURVEILLANCE FOR VETERINARY DRUG RESIDUES

2.6.1 Antimicrobial agents

2.6.1.1 Antimicrobial activity

Antimicrobial inhibition tests, particularly the FPT, are used widely to detect residual antimicrobial activity in animal tissues. This test was used by the WPVR in its early surveillance for antimicrobial residues and is now used for the same purpose within the National Surveillance Scheme (section 2.4.2). From 1977 to 1979 a survey was carried out on samples of UK-produced meat and imported meat and offal. The UK-produced material was obtained from abattoirs which were licensed to export their products. Imported material was obtained from ports and wholesale markets. The results (Table 2.5),

Table 2.5. Survey of antimicrobial activity in UK-produced and imported tissue samples: 1977 to 1979

Source	Animal type	Sample	Number of samples tested	Number of samples giving a positive result
UK produced	Cattle	Meat	2076	None
	Pigs	Meat	1862	1
	Sheep	Meat	1244	None
	Horses	Meat	260	1
Imported	Cattle	Meat	100	None
	Calves	Meat	106	None
	Sheep	Meat	525	None
	Unknown	Offal†	235	None

† Heart (64 samples), kidney (94 samples), liver (66 samples) and sweetbread (pancreas or thymus, 11 samples).

suggested a very low incidence of antimicrobial residues. Neither of the two positive samples out of over 6000 tested could be confidently identified or quantified.

A similar study was started in 1980, in which only meat and kidney produced in Great Britain were sampled. The number of samples taken at each slaughterhouse was proportional to the throughput of animals. The results (Table 2.6) were positive for about 1% of animals tested. Fewer positive results were found in sheep samples than in tissue samples from other animals, possibly reflecting the less intensive systems of agriculture used for sheep. The incidence of positive results was almost constant throughout the duration of this survey and did not exceed 3% during any three-month period. A higher incidence of positives was detected in kidney samples than in meat. This was either due to concentrations of some antimicrobials being higher in kidney than in meat, or because of

Table 2.6. Survey of antimicrobial activity in meat and kidney samples from quadrupeds slaughtered in Great Britain: 1980 to 1983

Animal type	Numbers of animals from which samples were tested	Numbers of animals from which samples gave a positive result
Cattle	815	10
Calves	891	16
Pigs	853	11
Sheep	828	4

substances naturally present in the kidney having some antimicrobial activity. When the positive samples were re-examined by electrophoresis it was not possible to identify or quantify the substances present.

The failure to identify or quantify the positive results found in the 1977 to 1979 survey and the 1980 and 1983 survey, caused the WPVR to question the validity of using antimicrobial inhibition tests in surveillance work (see section 2.5.5). This led the WPVR to place greater emphasis on the development of quantitative chemical methods.

In the USA microbiological inhibition tests have also been used to detect antimicrobial residues at or above tolerance levels in animal tissues. During 1985 about 1.4% of 3134 kidney samples from cattle, pigs, sheep, horses and goats gave positive results. However, when muscle samples from the same animals were tested the majority were negative. The US Food Safety and Inspection Service (FSIS) found that when violations did occur they were frequently in mature or culled animals sold through intermediate markets or in newborn calves that were not retained for herd replacement. Such animals were often medicated and sent for slaughter before sufficient time had elapsed for the drugs to be eliminated from the animal. The FSIS introduced tests that can be applied to animals at the slaughterhouse. This has resulted in considerable reductions in the number of residue violations occurring in the USA.

2.6.1.2 Sulphonamide compounds

The emphasis by the WPVR on developing quantitative chemical methods for detecting antimicrobial residues (section 2.5.5) led to a small survey for sulphonamide residues in kidney samples by gas chromatography, and gas chromatography linked to mass spectrometry, in 1980 to 1983. The analytical method was capable of detecting 15 sulphonamide compounds at a limit of detection of 0.01 mg/kg. The results of a survey of samples obtained from British slaughterhouses using this approach are given in Table 2.7. Sulphadimidine was the only sulphonamide detected in this survey. It was found in 21% of pig kidney samples at concentrations generally higher than those detected in other animals. The WPVR, concentrating its attention on pig kidney samples, conducted a further small survey of 50 pig kidney samples in 1985. In this survey tandem mass spectrometry was used in which five sulphonamide compounds could be detected with a limit of detection of 0.05 mg/kg. Samples for this survey were obtained from British slaughterhouses in a random manner via the National Surveillance Scheme (section 2.4.2). Sulphadimidine was again the only sulphonamide compound detected in 14 out of the 50 samples. The data from the 1980 to 1983 survey were seen by the Committee on Toxicity of Chemicals in Food, Consumer Products and the Environment, and the Veterinary Products Committee (VPC) (see Chapter 9, Fig. 9.2). The VPC defined an MRL of 0.1 mg/kg for sulphonamides in edible animal products. This concentration was exceeded in 20 (12%) of the pig kidney samples in the 1980 to 1983 survey, and 7 (14%) of the pig kidney samples in the 1985 survey. As a result of a publicity programme targeted at the producer and the recent MRL regulations (section 2.2.3) the incidence of sulphonamides in pig tissues is now (1992) about 5%.

A similar position regarding sulphadimidine in pigs' tissues has been reported from other countries. The United States FSIS found that 5.3% of pigs' liver samples tested in 1985 had residues of sulphadimidine above 0.1 mg/kg. In a small survey conducted in

Table 2.7. Residues of sulphonamides in kidney samples from quadrupeds slaughtered in Great Britain: 1980 to 1983 (mg/kg)

	Number of animals from which		
Animal type	samples were analysed	samples contained sulphonamide residues	Residue concentrations
Cattle	87	3	0.01 to 0.27 (mean: 0.1)
Calves	89	3	0.09 to 0.36 (mean: 0.18)
Pigs	163	34	0.01 to 6.0 (mean: 1.5)
Sheep	96	None	—

West Germany from 1983 to 1985, 2 out of 24 samples of offal (mainly pigs' kidneys) and 3 out of 32 samples of calf and pig meat contained residues of sulphadimidine. Both of the offal samples contained sulphadimidine residues above 0.1 mg/kg. Surveillance for sulphonamide residues, particularly in pigs' tissues is continuing. In the UK samples of imported pigs' offal are also being analysed for residues of sulphonamides. In 1989, 63 imported pig kidney samples were analysed and no residues were detected. Sulphonamides were not detected in any of the 75 samples of imported bovine kidney which were also analysed in 1989.

2.6.1.3 Chloramphenicol

Because of concerns about potential toxic effects of chloramphenicol the WPVR commissioned the development of a gas chromatographic analytical method to detect residues down to 0.01 mg/kg. In 1984 and 1985 kidney samples from 79 cattle, 75 calves, 75 pigs and 75 sheep were obtained via the SVS National Surveillance Scheme. Although no residues of chloramphenicol were detected in any of those samples the WPVR continued its surveillance. In 1989, 58 samples of imported bovine kidney and 94 samples of imported pig kidney were analysed. No residues of chloramphenicol were detected.

In the USA the FSIS tested over 1000 samples of calves' muscle each year between 1982 and 1984, and nearly 300 samples in 1985. Chloramphenicol was detected in four samples in 1982, five samples in 1983 and one sample in 1985. The United States FDA does not permit the use of chloramphenicol so these results indicate illegal use. No residues were found in an FSIS survey of other livestock and poultry groups.

In West Germany residues of chloramphenicol were detected in 11 out of 80 samples of trout analysed between 1983 and 1985. In an Italian study of retail eggs, residues of chloramphenicol were detected in 84 out of 300 samples during 1986. Concentrations ranged from 0.01 to 0.29 mg/kg.

2.6.1.4 Tetracycline compounds

As a result of evidence suggesting that oxytetracycline may be used in fish farming in the UK, a survey of farmed trout was conducted by the WPVR during July and August 1984. Fifty-four samples of farmed trout, each containing eight to ten fish, were obtained from fish farms, wholesale and retail outlets and analysed using HPLC for tetracycline, chlortetracycline and oxytetracycline residues. The limit of detection of this method was 0.005 mg/kg for each tetracycline compound. Oxytetracycline was detected in seven samples, at 0.008 to 0.04 mg/kg. One sample contained tetracycline at 0.03 mg/kg but no residue of chlortetracycline was detected. These results suggest that residues of tetracycline compounds were present in about 16% of the farmed trout sampled.

The risk of bacterial infections in fish farms is greatest during the warmer months of the year and so it is likely that the use of antimicrobials would have been at its maximum at the time these samples were taken. However, fish infections can occur at other times of the year. Antimicrobials which are applied during colder periods will remain in the fish longer because fish metabolic rates are slower.

In 1989, 96 samples of veal fillets were analysed for residues of tetracyclines. Of these, 11 contained residues above the limit of detection of 0.01 mg/kg, but below the limit of detection of the confirmatory technique (0.05 mg/kg). A number of the samples were not of UK origin.

2.6.1.5 Furazolidone

The samples of farmed trout collected for analysis for tetracyclines (section 2.6.1.4) were also analysed for residues of furazolidone as there was evidence that this agent was being used in fish farming in the UK. An HPLC method with a limit of detection of 0.02 mg/kg was used to analyse 50 batched samples, each containing eight to ten fish. No residues of furazolidone were detected but it has since been shown that this compound is rapidly metabolized (section 2.3.1).

In an Italian study of eggs, residues of furazolidone were detected in 275 (92%) of 300 samples taken during 1986. Concentrations ranged from 0.01 to 0.28 mg/kg.

2.6.2 Anabolic agents

2.6.2.1 Stilbenes

Surveillance for residues of stilbene compounds began before the use of these compounds was prohibited in the UK in 1982. Samples of kidney and liver from 372 calves were analysed by immunoassay from December 1980. Wherever possible, when stilbene residues were detected, samples of muscle from the same animal were also analysed. Work continued and, in 1982, a more rapid radioimmunoassay using bile became available. Between March 1982 and May 1986 samples of bile from 4839 calves, adult cattle, pigs and sheep were analysed. Whenever possible, where stilbene residues were detected in bile, samples of kidney, liver and/or muscle were analysed. The limits of detection varied from 0.0005 to 0.001 mg/l in bile and from 0.0001 to 0.0002 mg/kg in tissues. The results of this surveillance from 1980 onwards (Table 2.8) indicate that residues of stilbenes occurred infrequently and, with the exception of one sample of pig bile, at very low concentrations. There was a marked decrease in incidence of positives in the years

Table 2.8. Stilbene residues in samples from quadrupeds slaughtered in Great Britain, 1980 to May 1986

	Numbers of animals from which		Concentration (mg/kg or mg/l) of stilbene		
Animal type	samples were analysed	samples contained stilbene residues	Sample	Range (number of samples)	Mean
Calves	1401	30	Bile	0.002 to 0.1 (3)	—
			Kidney	0.0002 to 0.003 (18)	0.002
			Liver	0.0003 to 0.002 (9)	0.001
			Muscle	0.0003 to 0.001 (12)	0.0004
Cattle	1228	18	Bile	0.001 to 0.03 (18)	0.009
			Kidney	0.001 (1)	—
			Liver	0.0003, 0.0004 (2)	—
			Muscle	0.0001 (1)	—
Pigs	1307	18	Bile	0.002 to 4 (18)	0.2
			Muscle	0.0001 (2)	—
Sheep	903	1	Bile	0.004 (1)	—

following the prohibition on the use of these compounds (Fig. 9.1, Chapter 9). The incidence of stilbene residues in quadrupeds tested in 1986 was 0.15%. Although this proportion was very small it may have represented illegal use of these compounds in many thousands of animals. No stilbene residues have been detected in quadrupeds or poultry since 1986. Surveillance for stilbene residues is continuing and if residues are found the sample will be traced back to the farm of origin. If further evidence of use can be found on the farm then the farmer may be prosecuted.

Prior to the prohibition of stilbene compounds in the UK in 1982 diethylstilboestrol and hexoestrol were licensed for the caponization of poultry. The neck was the usual implant site. Samples of leg muscle and neck from 42 birds were obtained from retail outlets in England and Wales between December 1983 and January 1984. The limit of detection of the radioimmunoassay used was about 0.001 mg/kg. Stilbene residues were detected in samples from two birds but their presence was not confirmed by gas chromatography–mass spectrometry (GC/MS).

During 1984 samples of imported kidney from 50 veal calves (Netherlands) and livers from 50 cattle (Republic of Ireland) were analysed for stilbene residues. No residues of stilbene compounds were detected. The use of stilbenes is also not permitted in the USA and the FSIS conducts surveillance to detect residues of these agents in cattle and calves. No residues above the limit of detection of the GC/MS method, of 0.00025 mg/kg, were detected during 1986.

2.6.2.2 Trenbolone and related compounds

Trenbolone acetate is rapidly metabolized in cattle to three compounds: trenbolone (the 17β-hydroxy metabolite), epitrenbolone (the 17α-hydroxy metabolite) and trendione (the 17-keto metabolite) (Fig. 2.7). Early research work was primarily limited to trenbolone, and initial surveillance for residues by the WPVR was targeted on this compound. Between 1980 and 1983, 901 samples of calf plasma and 745 samples of cattle liver were analysed for trenbolone by radioimmunoassay. (The limits of detection were 0.0002 mg/l for plasma and 0.0003 mg/kg for liver.) Trenbolone residues were detected in two samples of calf plasma, at 0.0002 mg/l in both cases. Residues were detected in 16 samples of cattle liver from 0.0003 to 0.003 mg/kg (mean: 0.0006 mg/kg). However, during the period of this surveillance new information showed that the major metabolite of trenbolone acetate is epitrenbolone. As a consequence of this the WPVR modified its surveillance to include both trenbolone and epitrenbolone. During 1984, 50 samples of imported calf kidney from the Netherlands, together with 50 samples of cattle liver (Republic of Ireland), were analysed for these residues. No residues of either trenbolone or epitrenbolone were detected in any of these samples above the limit of detectioin of 0.0002 mg/kg for each isomer.

Under the NSS, a total of 2568 bovine bile samples from slaughterhouses in Great Britain were analysed between 1986 and 1990. Residues of trenbolone were confirmed in 11 samples. A total of 248 samples of bovine bile from slaughterhouses in Northern Ireland were screened for trenbolone between 1989 and 1990. Of these, 11 contained residues above the limit of detection, of the enzyme-linked immunosorbent assay used, of 0.0014 mg/l.

In a West German survey of 62 imported calf muscle samples carried out between 1983 and 1985, residues of trenbolone and trenbolone acetate were detected in seven samples.

2.6.2.3 Zeranol

A survey of zeranol residues in samples from British cattle, calves and sheep was carried out between 1983 and 1984. Zeranol was detected in 31% of bile samples from cattle and 4% of bile samples from both calves and sheep (Table 2.9). Whenever possible muscle samples from animals shown by analysis of bile samples to have been treated with zeranol, were also analysed. The failure to detect residues of zeranol in some samples of muscle from animals with contaminated bile reflects the lower residue concentrations of zeranol in muscle than in bile. Zeranol concentrations in meat ranged from 0.0003 to 0.0007 mg/kg, whereas concentrations in bile were between 0.001 and 0.04 mg/l.

The analytical method used in this survey involved the use of a monoclonal antibody in a radioimmunoassay. These antibodies are very specific and were used to reduce interferences from the naturally occurring fungal metabolite zearalenone (Chapter 5) and its metabolic product zearalenol. The limits of detection were 0.001 mg/l in cattle and calf bile, 0.006 mg/l in sheep bile and 0.0003 mg/kg in muscle.

Via the NSS a total of 3921 bovine and ovine bile samples from slaughterhouses in Great Britain were analysed between 1987 and 1990. Residues of zeranol were confirmed in 92 samples with a limit of detection of 0.002 mg/l. A total of 299 samples of cattle and sheep bile from slaughterhouses in Northern Ireland were screened for zeranol between

Table 2.9. Zeranol residues in samples from quadrupeds slaughtered in Great Britain in 1983 to 1984

		Numbers of animals from which:	
Animal type	Sample	Samples were analysed	Samples contained zeranol residues
Cattle	Bile	198	61
	Muscle	41	19
Calves	Bile	152	6
	Muscle	4	2
Sheep	Bile	58	2
	Muscle	2	0

1989 and 1990. No residues levels above the limit of detection of the enzyme-linked immunosorbent assay used (0.0014 mg/l), were detected.

A survey in West Germany detected residues in six out of 62 samples of imported calf meat examined; one out of 51 samples of imported calves' kidney was found to contain zeranol residues at a concentration of 0.01 mg/kg.

2.6.2.4 Natural hormones

In most cases the proper use of naturally occurring anabolic hormones in animal production should not measurably increase the concentrations of these compounds above those found naturally in animal products. However, the WPVR had noted the results of research which suggested that the use of progesterone in male calves could increase the concentration of this compound in their fat, and that testosterone concentrations in the plasma of female calves could also be increased by use of this androgen (section 2.3.2). Samples of fat from 48 male calves and plasma from 152 female calves were obtained from British slaughterhouses in 1980 and 1981. These were analysed by a radioimmunoassay which used commercially available assay materials and gave limits of detection of 0.0001 mg/kg for progesterone and 0.00005 mg/l for testosterone. Progesterone was detected in all fat samples at 0.0009 to 0.005 mg/kg, and testosterone was present in 108 plasma samples from 0.00005 to 0.0009 mg/l. There was no evidence that the use of these two naturally occurring anabolic agents could significantly increase their concentrations in fat or plasma above the naturally occurring concentrations, of about 0.006 mg/kg for progesterone in fat from male calves and about 0.00006 mg/l for testosterone in plasma from female calves.

A survey of hormonal growth promoter implants found in cattle has compared the pattern of usage of such implants before and after their prohibition on 1 December 1986. Samples of ears were taken at random from a slaughterhouse that received cattle from Scotland and Northern England. Implants were sought by palpation of the ears after slaughter. Table 2.10 shows the patterns of use in 1984/5 and in 1987. The number of

Table 2.10. Patterns of use of hormonal implants in steers in Scotland and Northern England

	Numbers		Chemical composition of implants (%)			
Period of study	examined	with implants (%)	Trenbolone	Oestradiol	Zeranol	Others
July 1984 to June 1985	1159	93 (8)	38	47	6	8
March 1987 to September 1987	469	60 (13)	14	83	0	3

implants of zeranol detected was probably an underestimate because their pellets are small and difficult to detect. The results of this study suggest that the use of implants in steers did not decrease as a result of the prohibition on their use. The main trend was a shift away from the use of synthetic hormones trenbolone and zeranol, the residues of which are readily detectable, to the natural hormones (notably oestradiol) the use of which are not easily discernible from the endogenous hormone. However under the NSS, a total of 4454 bovine serum samples from slaughterhouses in Great Britain were analysed between 1987 and 1990. Residues of natural hormones (oestradiol, progesterone and testosterone) above threshold values were detected in 40 samples. A total of 478 samples of bovine serum from slaughterhouses in Northern Ireland were screened for natural hormones between 1989 and 1990. Residues above threshold values were detected in seven samples.

2.6.3 Anthelmintic agents

2.6.3.1 Levamisole

Levamisole is no longer widely used as an anthelmintic for cattle in the UK. Samples of liver from 50 cattle were obtained from British slaughterhouses in 1984. They were analysed using the US Department of Agriculture method which has a limit of detection of 0.1 mg/kg. No residues of levamisole were detected in this survey. In 1986 to 1987, 106 samples of milk from UK farms were analysed for residues of levamisole using gas chromatography. No residues were detected above the limit of detection of 0.05 mg/l.

The US FSIS analysed 1035 liver samples from sheep, pigs and cattle for levamisole during 1985. Residues were detected in three liver samples (two pig, one cow). The residue of levamisole in the cow's liver was above the US tolerance level.

2.6.3.2 Benzimidazole compounds

The WPVR supported the development of a multi-residue HPLC analytical method for the detection of albendazole, cambendazole, fenbendazole, oxfendazole and thiabendazole (Fig. 2.8) and their sulphoxide and sulphone metabolites in liver and meat. The method was tested on tissues from animals which had been treated with these compounds. This demonstrated that the method worked and provided further evidence that residue concentrations of these benzimidazole compounds were higher in liver than in other tissues. A survey of liver samples from 50 sheep, 25 cattle and 25 calves obtained from British slaughterhouses was conducted during 1984 and 1985. No residues of these compounds or their metabolites were detected in any of the samples, however.

Samples of imported bovine, ovine and porcine offal were analysed in 1989 and 1990. In 1990, four samples of sheep liver imported from New Zealand were found to contain residues of oxfendazole, at concentrations up to 0.59 mg/kg. The MRL for this residue in liver was 1 mg/kg. The US FSIS has also developed an HPLC multi-residue method for the detection of benzimidazole residues. This method can detect thiabendazole (and its 5-hydroxy metabolite), fenbendazole, oxfendazole and mebendazole. Surveillance of cattle and sheep during 1985 detected no residues of these compounds in 365 samples.

2.6.3.3 Ivermectin

Ivermectin is widely and increasingly used as an anthelmintic in the UK and so WPVR surveillance includes this compound. Between 1989 and 1990, 256 bovine kidney, 134 pig kidney and 94 veal fillets samples were analysed for residues of ivermectin. No residues were detected using a method with a limit of detection of 0.002 mg/kg. It is also gaining widespread use in the USA. The US FSIS monitored for residues of ivermectin, using an HPLC method, in 231 samples of pigs' tissues during 1984. Only one residue, in liver, above the US tolerance level was detected. In 1985 surveillance was extended to include cattle and sheep. Out of 748 samples analysed, only one residue, in pig's liver, was detected above the US tolerance level.

2.7 POTENTIAL HUMAN HEALTH EFFECTS OF VETERINARY DRUG RESIDUES IN FOOD

Surveillance for veterinary drug residues has shown that traces of these compounds can, and sometimes do, arise in food. Although all of these compounds have potent biological activities in order to be effective in use, it is necessary to ensure that any residual activity in a food product is not harmful to the consumer. The use of veterinary medicines will inevitably lead to the presence of some residues in food. The purpose of toxicological safety evaluation is to determine at what concentration the effect of residues of a particular compound on human health becomes a cause for concern. To this end dose-response relationships need to be established. Such relationships can be used to determine the concentration of a drug at which the risks to human health become acceptable and are outweighed by the benefits from the use of the drug. This is essentially the process involved in the setting of Acceptable Daily Intakes (ADIs) and recommending withdrawal periods described in section 2.2. The subjects of toxicity and risk-benefit analysis are considerable disciplines in themselves and it is beyond the scope of this book to discuss them in detail.

The risks from the presence of veterinary drug residues in food may be primarily limited to a very small minority of susceptible individuals. The presence of residues tends to be isolated to relatively infrequent occurrences. However, it is the duty of those responsible for monitoring the safety of the food supply to ensure that adequate surveillance is conducted and controls applied, to protect the health of all consumers.

2.8 CURRENT ISSUES RELATING TO VETERINARY DRUG RESIDUES IN THE UK

2.8.1 Introduction

Fifteen years of surveillance for veterinary drug residues in food in the UK has presented a generally reassuring picture. The residues of veterinary drugs that have been detected have been at very low concentrations and generally at low frequencies. Where repeated surveillance has been carried out downward trends have often been observed in the incidence of residues detected. However, some contamination 'hot spots' have been revealed and these then received close examination. Furthermore new products are constantly being introduced into the market and these need effective evaluation of the risks from residues. This is done by the licensing processes described in section 2.2. The UK imports a significant amount of animal-derived food. The likelihood of residues occurring in this part of the food supply must also be considered to ensure the quality is that expected by the UK consumer.

2.8.2 Antimicrobial residues

Microbiological inhibition tests have been found to be an unsatisfactory means of detecting residues of antimicrobial agents in meat. As a consequence, considerable resources have been devoted to developing specific and sensitive chemical methods of analysis. Surveillance using these new methods has revealed the presence of residues which may well have remained undetected by previous biological assays.

Particular effort has been devoted to controlling the incidence of residues of sulphadimidine in pig meat. In surveys conducted between 1980 and 1985, residues above the then maximum acceptable residue concentration of 0.2 mg/kg were detected in 12% to 15% of samples analysed. Subsequent work has shown a steady decline in the number of samples containing residues in excess of the MRL (0.1 mg/kg) to about 5%.

Sulphadimidine mixed with feed is given to pigs, to control respiratory infections. The VPC recommends that the drug should be withdrawn from the animal's diet at least ten days before the animal is taken for slaughter. The withdrawal period should allow sufficient sulphadimidine in the animal's body to be metabolized or excreted. Farmers find the application of such withdrawal periods difficult to administer in large pig units where all the animals are receiving their feed through a common, mechanized feeding system. It is widely believed that the presence of sulphonamide residues in pig meat reflects a failure of producers to observe withdrawal periods. However, there is evidence that powdered sulphonamides can adhere to the walls of mixing vessels. This means that non-medicated feeds mixed in a vessel after a medicated feed may become contaminated.

The UK Ministry of Agriculture, Fisheries and Food has approached pig farmers, the veterinary profession, and feed compounders to explain the problem and seek their co-operation in resolving it. The use of granular sulphadimidine formulations instead of powder may help to control cross-contamination in mixing vessels. Recently introduced UK legislation (S.I. [1991] No. 2843) will permit legal action to be taken against farmers who fail to observe withdrawal periods and produce pig carcasses containing sulphonamide residues in excess of the MRL. Similar legislation is to be introduced by the EC in the future.

An encouraging aspect has been the reaction of some major food retailers. Such retailers have a major influence in the market and several are now stipulating that no medicated feeds should be used for pigs, which they will purchase, during the 'finishing' period of pig production.

2.8.3 Anabolic hormone residues

The administration of the potent oestrogenic stilbene compounds was prohibited in the UK in 1982. However, in 1986 residues of stilbenes were detected in 0.15% of cattle, calves, pigs and sheep sampled in UK slaughterhouses. On 1 December 1986 the administration of all other anabolic hormones for growth promotion purposes was prohibited in the UK. However, a survey conducted at a slaughterhouse before and after the prohibition, showed no apparent decline in the proportion of steers (the principal animal group on which these drugs were used) with remnants of implants in their ears. The major change that had occurred was a shift from synthetic hormone implants to the natural hormone oestradiol. Anecdotal evidence from other EC Member States indicates that farmers may be using other means to avoid detection. Injections of oily suspensions of hormone 'cocktails' and deep injections of 'gel depot' hormones have been found which are intended to emulate the slow-release characteristics of the previously licensed implants. Such injections are apparently not confined to tissues such as the ears, which are normally discarded at slaughter. This uncontrolled and illegal use of anabolic hormones may result in injection sites containing very high concentrations of hormones entering the food supply. There is evidence of a growing 'black market' in such preparations in Europe. The majority of scrupulous farmers who avoid using such illegal products may believe themselves to be at an economic disadvantage when competing with those who do. The more prevalent this becomes, the more farmers may feel pressurized to join it. The unfortunate consequence of the hormones prohibition in the EC may be that consumers may be at greater risk from hormone residues now than they were when certain hormonal growth promoters were licensed and properly administered. However, conclusive evidence on the actual usage of illegal hormone preparations is lacking—indeed in the UK extensive food surveillance has found no evidence of residues that would cause harm to consumers.

2.8.4 Surveillance for residues of newly licensed veterinary medicines

Any application for a licence to market a new veterinary medicine undergoes a thorough evaluation by the Veterinary Products Committee (section 2.2.1). The licensing process ensures that if a product is used according to the conditions stated on the data sheet accompanying it, harmful residue concentrations will not be present in tissues of treated animals entering the food supply. This would produce a highly satisfactory situation as long as the conditions of use are always strictly observed. However, in practice experience has shown that in a small number of cases this does not occur and problems such as those with sulphonamides in pig meat can occur. Consequently surveillance is essential in order to provide an ultimate check on the effectiveness of the licensing procedure.

2.8.5 Veterinary drug residues in imported animal products

Surveillance for veterinary drug residues in animal products imported into the UK has indicated that the incidences and concentrations of residues in such food are in general similar to those in animals produced on UK farms. However, some reports of surveillance in other countries suggest that this may not always be the case. This may be due in part to licensing procedures differing from country to country or it may reflect less effective surveillance for residues than that conducted in the UK. The introduction of the free European market will mean that licensing and monitoring procedures will be harmonized throughout Europe.

2.8.6 Conclusions

There are a great many chemical substances licensed for use as veterinary medicines in the UK. Their use is essential and the majority of farm animals may receive a dose of some medication during their lifetime. Surveillance for residues in the UK, by the Working Party on Veterinary Residues in Animal Products, is directed towards substances which may present a potential hazard to the consumer if present at high enough concentrations in food. The results of this surveillance have shown that in this respect food safety in the UK is generally very good and with some minor exceptions, for example sulphonamide residues in pig meat, residues of veterinary drugs occur infrequently and at very low concentrations.

The implementation of legislation in the UK will ensure even more effective control over residues than has previously been possible. This may help to deter the small number of farmers who do not observe the recommended conditions of use for veterinary products and who are probably therefore responsible for the low incidence of contamination that is detected. Refinements in analytical methodology, to improve the speed and reliability of residues detection, should help to expand the coverage of the UK food surveillance programme.

2.9 SUMMARY

A large number of different veterinary products is licensed for use on farm animals in the UK and many other countries. The drugs include antimicrobials, anthelmintics, coccidiostats and other agents for therapeutic, prophylactic or growth promotion purposes. The use of anabolic growth promoters is prohibited throughout the EC. In the UK the licensing of veterinary products is carried out under the guidance of the Veterinary Products Committee and non-statutory surveillance is co-ordinated by the Working Party on Veterinary Residues in Animal Products. International committees have been established to standardize safety limits to facilitate trade between nations, and there is increasing harmonization of controls and surveillance in the EC.

Surveillance for residues is undertaken, and as evidence for potential human health risks emerges suitably sensitive analytical methods are developed. The results of surveillance have shown that food in the UK is generally very safe. Particular problems have been encountered with residues of sulphonamides in pig meat. Effective legislation to control problems such as this has been introduced into the UK and will, in the future, be implemented by the rest of the EC.

FURTHER READING

Food Surveillance Paper No. 22. HMSO, London. (1987)

Food Surveillance Paper No. 33. HMSO, London. (1992)

Rico, A. G. (ed.) *Drug residues in animals*. Academic Press, London. (1986)

Evaluation of certain veterinary drug residues in food. 32nd Report of the Joint FAO/WHO Expert Committee on Food Additives. World Health Organization, Geneva. (1988)

Cordle, M. K. *J. Animal Science*, **66**, 413–33. (1988)

The hormone issue 1980–1987. National Office of Animal Health, London. (1987)

Maddox, J. G. *The Veterinary Record*, **122**, 161. (1988)

Food Safety Act 1990. HMSO, London.

The Animals, Meat and Meat Products (Examination for Residues and Maximum Residue Limits) Regulations 1991 (Statutory Instrument [1991] No. 2843), HMSO, London.

Heitzman, R. J. (ed.) *Residues in food-producing animals and their products: reference materials and methods*. EC, Luxembourg. (1992)

Crosby, N. T. (ed.) *Determination of veterinary residues in food*. Ellis Horwood. (1991)

3

Dioxins and other environmental organic chemicals

D. H. Watson, Ministry of Agriculture, Fisheries and Food, R242, Ergon House, c/o Nobel House, 17 Smith Square, London SW1P 3JR, UK.

3.1 INTRODUCTION

Over the last two decades there has been growing interest in contamination of the food chain with organic chemicals that are either produced by industry or are by-products of industrial activity. Although compared with other areas described in this book the scientific study of this topic is still at an early stage, work on detecting such chemicals in food has followed on quickly from research showing that the substances are present in the environment. Pioneering work in the USA and Europe on environmental contamination with persistent chlorinated organic chemicals, in the 1960s and 1970s, led in the 1980s to the development of very sensitive methods of analysing food for chlorinated compounds and, indeed, several other types of industrial organic chemicals. There has also been intensive toxicological effort to identify and quantify the primary toxic effects of several industrial chemicals and chemical by-products, largely stimulated by efforts to improve worker safety. This too has helped to stimulate and establish the relatively new work on such substances in the food chain.

Table 3.1 lists some of these chemicals that can now be studied in food and raw materials used in food production. It is noticeable that even for the relatively few examples in this table the capability of current analytical methodology varies considerably. Detection limits vary by several orders of magnitude—from extremely low levels for dioxins and furans to a tenth of a part per million for some other chemicals. However, some extremely hazardous substances (e.g. benzene, which is now accepted as carcinogenic) are not yet readily detectable in food because analytical methods need to be developed fully. Like naturally occurring toxicants in food (Chapter 5) there is a pressing need for analytical methods to detect many of the industrial chemicals that might contaminate the food chain. But unlike work on natural toxicants there is a rapidly growing library of toxicology about many industrial chemicals. This has followed from the

Table 3.1. Some organic environmental chemicals that can currently be studied in the food chain

Chemicals	Currently achievable detection limits for these chemicals in food
Industrial chemicals:	
Benzene	? (not routinely detectable; suitably sensitive methods need to be developed)
Phthalates	*c*. 0.1 part per 10^6
Organophosphates, e.g. alkyl phosphates	*c*. 0.1 part per 10^6
Organochlorines, e.g. PCBs	*c*. 0.001 part per 10^6
Industrial by-products:	
Chlorinated dioxins	0.01 part per 10^{12}
Chlorinated furans	0.01 part per 10^{12}
Polynuclear aromatic hydrocarbons	*c*. 1 part per 10^{12}

intensive toxicological effort, to protect workers, noted above. Where analytical methods are available it is largely because of a cross-fertilization of effort from well-established areas of food contaminants work. For example, the steady development since the 1960s of methods of analysis for chlorinated pesticides led to the analysis of food for polychlorinated biphenyls (PCBs; Table 3.1) since PCBs were readily detectable by general methods used to analyse food for organochlorine pesticides. The analysis of food for chlorinated dioxins and furans at the very low levels at which they are found in food is a more recent development, and one which is an important precedent since it arose from interest in environmental contamination rather than because of cross-fertilization of scientific methodology from an established area of food chemistry. Although dioxins were detectable some years ago at much less sensitivity in some pesticides, it was environmental interest that led to their study at very low levels in the food chain. Growing analytical capability has provided a broader scope for those food chemists who are interested in the many organic chemicals that might contaminate the environment and the food chain.

To sustain this relatively new area of work a systematic process is needed for selecting chemicals to study. The following criteria are probably the key ones in the selection process:

— *Production volume* Small volumes of production of a given chemical should not necessarily eliminate it from study, but since there are as many as 50 000 different chemicals produced by industry worldwide it is important to exclude low volume chemicals where there is evidence that they are not toxic or environmentally persistent. Any threshold production volume could be used to reduce the list of target chemicals to be studied, but given the number of chemicals manufactured in most developed countries a minimum annual production of 1 tonne in a given country would probably be reasonable.

— *Pattern of usage* This is important because it helps to define the proximity of each chemical to the food chain. Clearly greater priority should be given to chemicals used on farms or in food factories provided that the other factors noted below and above do not put the chemical out of the reckoning.

— *Potential for release into the environment* Using highly toxic chemicals under carefully controlled containment should eliminate them from any possible contamination of the food chain. Thus, in theory at least, it is the less toxic, less contained chemical usages that might lead to food contamination. However, in practice it is very difficult to judge the probability of chemical release and this is a factor where available evidence is much more important than theory.

— *Likelihood of entering the food chain* Closely linked to pattern of usage, this factor must be assessed on the basis of observed data rather than by a theoretical approach. Although several mathematical models have been developed to estimate the flows of chemicals in the environment, and hence into the food chain, these models still need considerable data for validation. For organic chemicals entering the food chain the data are too few to apply precise mathematical models at present.

— *Persistence in the food chain* Although this is clearly a key factor there is very little known about it for most industrial chemicals and by-products. Not surprisingly there is relatively little work on mathematical modelling (see above) although it is known that some groups of chemicals, notably organochlorines, can persist for many years, for example in the body fat of farm animals. More could be done, perhaps, to extrapolate from the extensive and growing collection of experimental data on organic chemicals used in agriculture. This could help with information about the persistence of chemicals that are structurally similar to industrial chemicals.

— *Toxicity* Studies on worker and environmental safety have created a lot of information about the toxicity of industrial chemicals. However, less is usually known about the toxicology of industrial by-products such as dioxins. What is needed is information about what levels of exposure are unsafe—most chemicals can be shown to have some toxic effect if fed to experimental animals at high enough levels. Unique amongst the factors noted here, enough is known about the toxicology of some industrial chemicals for the researcher to discriminate between those chemicals that are worth studying in food and those that are not.

As this area of science develops it should be possible to select chemicals that should be looked for in the food chain by considering sound evidence in each of the above areas. At present the above factors provide a useful perspective in the prioritization process where evidence is available. Thus the range of organic chemical environmental contaminants that is currently studied in food is limited to a relatively few substances compared with the several thousand chemicals that might be studied were there evidence that they could contaminate food. Substances that are currently being studied in food or related materials are discussed in the next three sections of this chapter.

3.2 DIOXINS AND FURANS

Polychlorinated dibenzo-*p*-dioxins and polychlorinated dibenzofurans ('dioxins' and 'furans') are ubiquitous environmental contaminants. Their complex structures (Fig. 3.1) are apparently highly resistant to biological degradation. However, the degree of chlorination, which is from one to eight chlorine atoms per molecule, seems to determine the acute toxicity of the molecules. Four chlorine atoms at positions 2, 3, 7 and 8 leads to greatest acute toxicity. Probably too little is known about the toxicology of individual dioxins and furans for there to be any sound basis at present for correlating the variations in number or position of chlorine atoms on the molecules, with the chronic toxicities of the different molecules.

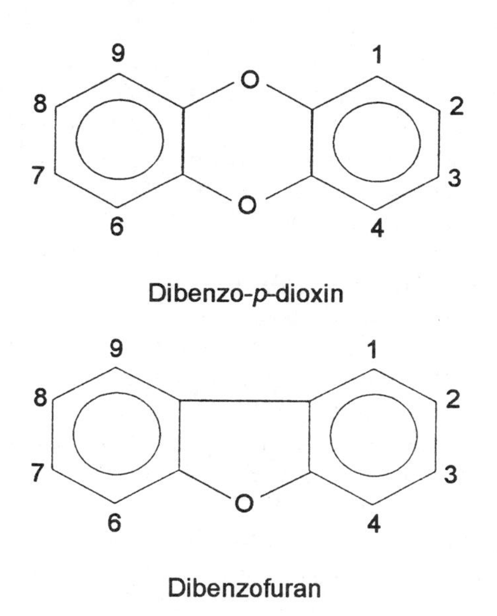

Fig. 3.1. Structures of dibenzo-*p*-dioxin and dibenzofuran. (Chlorine atoms can be attached to the numbered carbon atoms.)

Study of these substances is made particularly complex by the large numbers of different substances involved. There are 75 possible isomers of polychlorinated *p*-dibenzodioxin (PCDD) and 135 such isomers of polychlorinated dibenzofuran (PCDF). To aid the reporting of results a system of toxic equivalency has been developed whereby the total rather than individual amounts of PCDDs and PCDFs are reported, using weighting factors for each isomer to combine the individual amounts of isomers. Several ways of doing this have been developed. The most widely recognized one is the system proposed by the NATO Committee on Challenges to Modern Society, the International Toxic Equivalent Factor system. This concentrates on 17 tetrachloro-, pentachloro-, hexachloro-, heptachloro- and octachloro- derivatives of dibenzo-*p*-dioxin and dibenzofuran. These substances all have chlorine atoms at the 2, 3, 7 and 8 positions in the molecules. A weighting factor is assigned to each of these seventeen isomers. The other PCDDs and PCDFs are assumed to have relatively little biological activity and are given no weighting. The weighting factor for each isomer is multiplied by the amount of that isomer present in food and the resulting multiples are added together to produce a toxic equivalent. In this way the amounts of the different PCDDs and PCDFs are converted to one figure which is a measure of the toxicological impact of the detected mixture of PCDDs and PCDFs. Similar systems to the NATO one have been developed by a Nordic Expert Group, the German Federal Health Office and the US Environmental Protection Agency. These systems tend to give lower total figures for toxic equivalents as they use lower toxicity equivalencies for one or more of the pentachloro-PCDDs or PCDFs.

With the toxic equivalency approach the reporting of surveys for PCDDs and PCDFs in food is simplified, allowing the correlation of different surveys from country to country. It has also helped in the comparison of survey results with a recently established tolerable daily intake (TDI). This TDI, of 10 pg/kg bodyweight per day for 2,3,7,8-tetrachlorodibenzo-*p*-dioxin (TCDD), was established by an expert group convened by the World Health Organization (WHO) Regional Office for Europe. Taking into account toxic equivalency of the many PCDFs and PCDDs, this TDI can be regarded as 10 pg TCDD equivalents/kg bodyweight per day or seven times this for a time-weighted average tolerable intake per week. These are very useful parameters with which to compare toxic equivalents of PCDDs and PCDFs found in food surveillance.

Surveys for these substances in food have been carried out for only a few years. This is mainly because reliable, sensitive methods for analysing food for them have been developed only recently. The methods need to be capable of analysing at levels in the order of 0.01 ng/kg (1 part in 10^{14}). At this degree of sensitivity and with the complex mixtures of PCDDs and PCDFs that can be present, very strict criteria must be applied in carrying out and assessing the analytical work involved. The key criteria as defined by the Sub-Group on Dioxins (of the UK Steering Group on Chemical Aspects of Food Surveillance's Working Party on Organic Environmental Contaminants in Food—Fig. 9.2) are listed in Table 3.2. Applying these criteria should ensure that analytical results are properly representative of the sample that has been analysed (see Chapter 11, section 11.3 for a discussion of representative sampling). Survey results need to be very carefully assessed before they can be accepted and this must take account of how the analytical methods were applied, as well as the representativeness of the samples of food of the general supply. Because analysis for PCDDs and PCDFs in food is a particularly complex and hence expensive and time consuming process, particular care needs to be taken that sufficient samples are analysed in each survey.

Given all these difficulties it is not surprising that only recently have major surveys of PCDDs and PCDFs in food become a viable prospect. So far surveillance has been mainly of one type of food from a specific locale where there was suspicion of above-average contamination with PCDDs and PCDFs. These targeted surveys cannot provide information about general dietary exposure. Indeed their results can easily be misinterpreted if they are taken as a measure of general exposure. However, there have been estimates of total dietary exposure based on broader surveys in the UK, Germany and Canada. Dietary intakes of PCDDs and PCDFs estimated from these studies are as follows:

Table 3.2. Key criteria in laboratory analysis for PCDDs and PCDFs

General

Adequately documented method;
correct use of internal standards; and
specification of limits of detection.

Validation

Extraction efficiency;
blanks—average and repeatability;
recovery—average and repeatability;
selection of appropriate GC column;
selection of MS resolution and other conditions;
choice of monitored ions and cycle time for selected ion monitoring;
repeatability of relative response factors and isotope ratios; and
absence of suppression effects.

Quality control

Adequate blanks;
check sample extraction;
recovery of internal standards;
chromatographic separation;
mass spectrometer calibration and tuning; and
mass spectrometer sensitivity.

Data acceptance

Validated and quality controlled procedures;
criteria for identification as PCDD/PCDF;
criteria for assignment as specific isomer; and
criteria for acceptance of quantification.

— from UK work covering a range of 14 major components of the diet: 2.1 pg toxic equivalents (TEQ) per kg bodyweight (bw) per day for a 60 kg person (where TEQ is the 2,3,7,8-tetrachlorodibenzo-*p*-dioxin toxic equivalent);
— from two German studies of meat, milk and its products, eggs, fish, vegetable oil, vegetables and fruit: 1.6 and 3.3 pg TEQ/kg bw/day for a 60 kg person;
— from a Canadian study of ten main dietary components: 1.5 pg TEQ/kg bw/day for a 60 kg person.

The foods included in these estimates were of similar types and, given the analytical difficulties that can be experienced in such surveys, it is encouraging that the estimates of intake were also similar. Such estimates provide a context for more specific work to estimate dietary intakes of PCDDs and PCDFs in more detail, both for the general population and amongst sub-groups of the population that might be thought to be exposed to above-average levels of these substances in the diet. They can also be compared with the TDI noted above, even though further studies are probably needed to refine the intake estimates. Comparison with the TDI shows that intake is well below the TDI. The TDI is equivalent to 600 pg TEQ/day for a 60 kg person which compares with an estimated intake of 125 pg TEQ/day from the UK work.

Given these advances it should now be possible to study dietary contamination by PCDDs and PCDFs in more depth. There are several possible ways of doing this.

— General dietary and environmental surveillance for these substances may identify particular locales or types of food which are more likely to contain elevated levels of the compounds.
— It should be possible to study the transfer of PCDDs and PCDFs from known environmental sources through the food chain.
— Particular types of foodstuff might contribute more than average to dietary exposure to PCDDs and PCDFs.
— Consumers with particular dietary habits, for example the very young, the aged or the infirm, may have below-average or above-average intakes of these substances.

Although there are some indications that these strategies may be worth pursuing, there are not enough experimental data at present to identify any one of these as being more promising than the others.

A major problem in applying resources in this work is the shortage of evidence linking environmental sources and food contamination of PCDDs and PCDFs. Much of the data from environmental work point towards PCDDs and PCDFs being ubiquitous environmental contaminants, with direct evidence for above average levels in the environment being the main cause of food contamination in only a few well established cases (e.g. see work on contamination at some sites in Derbyshire reported in *Food Surveillance Paper No. 31*, HMSO, London).

The known sources of PCDDs and PCDFs are many, including the following:

— *Chemical manufacture* Chlorination of organic chemicals may lead to the formation of PCDDs and PCDFs as by-products. But where this has been detected, action has generally been taken to avoid formation of the compounds. Environmental contamination from this source may now be partly historical.
— *Bleaching processes* Use of chlorine to bleach wood pulp is an established source of PCDDs and PCDFs. However, this type of wood pulp treatment is now increasingly done using non-chlorine based processes, for example by using hydrogen peroxide. But continuing use of chlorine in bleaching other materials may also lead to PCDDs and PCDFs.
— *Combustion processes* It is thought by many that combustion is a major source of these substances. PCDDs and PCDFs have been found in exhaust gases from a wide variety of combustion processes—from cigarette smoke to emissions by fossil fuel power plants. Controls are being introduced on waste incinerators in the UK and several other countries.

The estimated relative contributions to tetrachlorodibenzo-*p*-dioxins in the environment from sources in the UK are shown in Fig. 3.2. The wide range of potential sources may well explain why PCDDs and PCDFs are ubiquitous, but it does not make it any easier choosing between possible ways of investigating their presence in food. However, it is believed that there are two main routes which PCDDs and PCDFs follow into the environment: via the atmosphere leading to deposition on soil, water and plants, and via solid or liquid waste with subsequent contamination of land (e.g. via sewage sludge). These findings should help to reduce the already extensive effort needed to trace back the contamination to its sources, although the persistence of PCDDs and PCDFs in the environment for many years means that historical as well as current sources need to be taken into account.

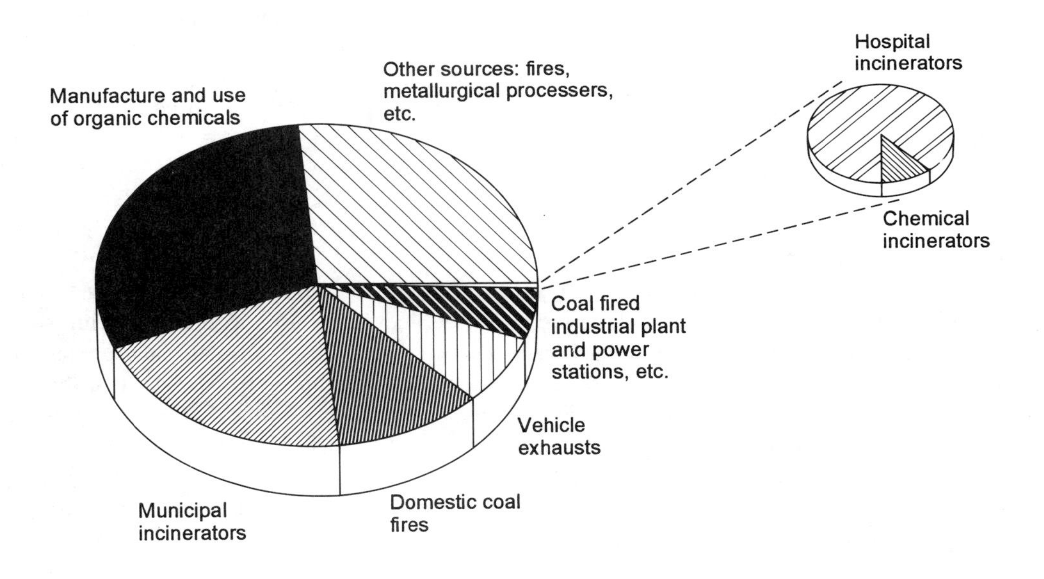

Fig. 3.2. Estimated relative contributions to total TCDD released into the environment from sources of dioxins and furans in the UK.

The recent advances described in this section should allow more extensive work on PCDDs and PCDFs in the food chain. It is unlikely that future work will be on a much greater scale than recent research unless quicker and less expensive methods of food analysis are developed for these substances. But the recent work described above does indicate that considerable advances in knowledge about these compounds can be made on the basis of a general picture of how PCDDs and PCDFs are produced, persist in the environment, and transfer into the food chain.

3.3 OTHER HALOGENATED ORGANIC CHEMICALS

Much of the effort on environmental chemicals that contaminate food has concentrated on a small range of chlorinated chemicals. In addition to recent work on PCDDs and PCDFs (section 3.2), there has been a lot of research since the 1960s on polychlorinated biphenyls (PCBs) and organochlorine pesticides. These two groups of environmentally persistent chemicals are detected by many group-specific methods of analysis for organochlorines in food, and so it is no coincidence that there are extensive data for both PCBs and organochlorine pesticides in the diet. There has also been some work on polybrominated biphenyls (PBBs). Although organochlorine pesticides have been studied in food mainly because of their use on crops and animals (Chapter 8), their approved uses have now become so restricted that for some of them at least they are now more likely to contaminate food as a result of their persisting in the environment rather than from their direct use on the food chain. The former pesticides now in this environmental category are hexachlorobenzene (HCB), DDT and its breakdown products, and arguably hexachlorocyclohexane (HCH) which is still used as a pesticide in some countries. There has also been some work on residues of *p*-dichlorobenzene in food, a disinfectant, and other industrial organochlorines such as 1,1,1-trichloroethene which is used as a dry cleaning fluid. Each of the above substances is reviewed below.

3.3.1 PCBs and PBBs

PCBs (Fig. 3.3) were once used in a wide variety of industrial applications, since they are chemically inert and have good dielectric properties. They were used as dielectric fluids in capacitors and in other electrical machinery, and in a wide variety of materials such as carbonless copy paper. However, as evidence grew that they are very persistent in the environment and that they might be toxic, their usage decreased. In the 1970s and early 1980s they were replaced by other substances such as alkyl-aryl phosphates (see section 3.5). The main potential sources are now persistent residues in the environment and PCBs in old industrial equipment, although controls over the disposal of equipment mean that the latter source is decreasingly likely to contaminate the food chain.

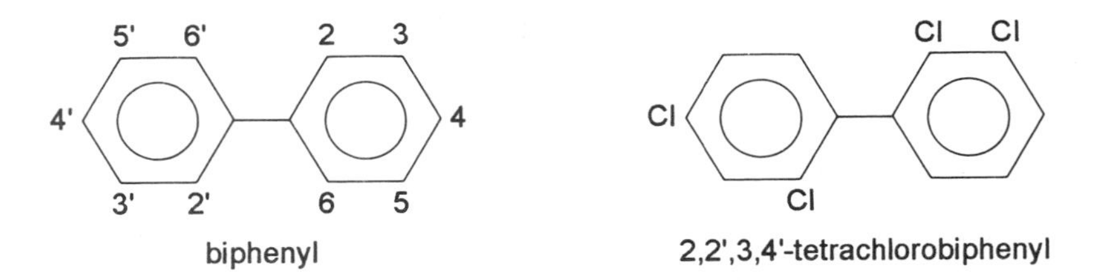

Fig. 3.3. Structure of PCBs. Chlorine atoms can be attached to the numbered carbon atoms of biphenyl as illustrated in the molecule on the right of the figure.

There have been many studies of PCBs in food and the environment over the past 20 years or so. The following are the main conclusions:

— PCBs are very persistent in both the environment and the food chain. Residues, which are usually in the parts per 10^6 range or less, can persist for many years since there is little or no biological degradation and no chemical breakdown.
— PCBs are hydrophobic being preferentially stored in body fat of man and other mammals. They are also partitioned in the environment onto particulate matter with a low water content.
— PCBs are therefore found in fatty foods and in the marine environment, particularly in fish with above-average fat contents. For the same reason they tend to be found at higher levels in fish liver than in muscle (Table 3.3).
— There is relatively little evidence that PCBs are toxic in the environment or to man. There are few reported incidents implicating PCBs in industrial illness and little clear evidence from *in vivo* studies in laboratory animals.

Table 3.3. PCBs in fish muscle and liver

		PCBs (mg/kg)	Number of samples
Cod	liver	4.1	73
	muscle	0.034	43
Dab	liver	0.91	50
	muscle	0.050	50
Flounder	liver	0.35	49
	muscle	0.040	49
Mackerel	liver	0.55	29
	muscle	0.13	29
Plaice	liver	0.40	68
	muscle	0.038	43
Sole	liver	0.30	50
	muscle	0.042	50
Whiting	liver	4.1	62
	muscle	0.031	62

Thus the main purpose of working on PCBs in food is to reduce exposure to them on general grounds of prudence because they are very persistent, rather than in the case of dioxins, because they are a potential toxic hazard. The clearest reason why there should be continuing work on PCBs in food comes from studies on human fat in which the decline in PCBs residues levels is probably very slow (Table 3.4). Not all of the residual PCBs in body fat may come from food, given their stability in the environment, but the food chain is an established source.

Much of the analytical data for PCBs in food were derived by analysis for these substances as a group. The results do not generally discriminate between different PCBs

Table 3.4. PCBs (mg/kg) in human fat from UK subjects over five years old

	1976 to 77 Sex and number of subjects		1982 to 83 Sex and number of subjects	
	Male 155	Female 81	Male 105	Female 82
Range	0.1 to 10	ND† to 1.5	0.1 to 6.9	ND to 2.2
Mean	0.8	0.6	1.0	0.8

† ND = not detected; limit of detection was 0.1 mg/kg.

isomers or group of isomers (congeners). In the past few years congener-specific methods of analysis have been developed which should allow distinctions to be made between the different groups of PCBs in food and the environment. The value of applying such methods to food has yet to be established however.

There has been much less work on polybrominated biphenyls (PBBs) although they were also used quite extensively by industry. There is evidence from the USA that PBBs are environmentally persistent and may well follow similar routes into and along the food chain as PCBs. However, their analysis in food is far from routine and there have been no extensive studies for PBBs in food. Further work is indicated given the parallels with PCBs.

3.3.2 HCB, DDT and HCH

These organochlorine pesticidal compounds are persistent in the environment. Although DDT is broken down to DDE, TDE and another isomer of DDT (*op′*-DDT; parent compound = *pp′*-DDT) in the environment, both the parent compound and its derivatives are persistent. Hexachlorobenzene (HCB) and hexachlorocyclohexane (HCH) are not significantly degraded to other chemicals. All of these organochlorines are now used very little if at all in the western world but their residues persist for some years in the environment. Like PCBs they are lipophilic with residues being found in human body fat (Table 3.5). However, their residual levels seem to decline more quickly than those of PCBs in human fat. There are probably too few data to define accurate environmental half-lives for HCB, HCH and DDT and its derivatives but they appear to be in the order of five to ten years.

There is very extensive information about residues of these substances in food but it is currently not possible to discriminate between residues that come from pesticide usage and environmental contamination. Clearly, above-average levels would be expected to arise from their use (or misuse) as pesticides but there are no hard and fast rules about what constitute 'above-average levels'. Residues in food are generally in the order of 0.001 parts per million, although levels of HCB are often a little lower than for the other substances. Acceptable daily intakes (ADIs) have been established for one of the three commonly found isomers of HCH (*gamma*-HCH; the other two isomers being *alpha* and

Table 3.5. DDT derivatives, HCB and HCH (mg/kg) in human fat (samples were the same as those in the survey in Table 3.5)

		1976 to 77		1982 to 83	
		Male	Female	Male	Female
β-HCH	Range	0.01 to 1.2	ND† to 1.1	ND to 0.77	ND to 0.81
	Mean	0.30	0.33	0.24	0.31
pp'-DDE		0.09 to 9.9	0.03 to 15	0.03 to 4.5	0.05 to 5.1
		2.1	2.1	1.3	1.4
pp'-DDT		ND to 0.91	ND to 2.4	ND to 0.42	ND to 0.53
		0.20	0.23	0.10	0.11
pp'-TDE		ND to 0.59	ND to 0.23	ND to 0.10	ND to 0.31
		0.03	0.03	0.01	0.02
Hexachlorobenzene		0.02 to 0.68	0.02 to 3.2	0.03 to 0.29	0.04 to 0.32
		0.18	0.22	0.11	0.13

† ND = not detected; limit of detection was 0.01 mg/kg.

beta) and for the sum of DDT, DDE and TDE. In the UK and many other western countries the estimated intakes of these substances are well below the ADIs. However, there is evidence that where crude mixtures of HCH isomers have been used, the most persistent (*beta*) isomer and the probably carcinogenic *alpha* isomer produce unacceptable residues in food and probably in the environment. The long-term impact of the persistent *beta* isomer on the food chain will continue to require continued surveillance, even if the use of crude HCH mixtures as a pesticide ceases in the less-developed countries where they are being used now or have been used recently.

This information points to the need for continuing surveillance and control of the food supply for these organochlorines even though their uses as pesticides are likely to decline to negligible levels in coming years. Fortunately methods of analysis for them in food are well developed, there are extensive databases built up over many years with which to compare new surveillance data, and there are ADIs for some if not all of the substances.

3.3.3 *p*-Dichlorobenzene and other industrial organochlorines

p-Dichlorobenzene (Table 1.2, Chapter 1) has been studied less extensively as a food contaminant than the other organochlorine compounds discussed above. However, it has been detected in human fat (Table 3.6), and its residues levels there are similar to those of other organochlorines (cf. Tables 3.4, 3.5 and 3.6). Although these residues could well have originated from mainly non-food sources, there are not the survey data for residues of this substance in food to confirm or refute this. *p*-Dichlorobenzene appears to fall into the same environmentally-persistent category as other organochlorine compounds and would be expected to give residues in the food chain.

Table 3.6. *p*-Dichlorobenzene (mg/kg) in human fat (samples were the same as those in the survey in Table 3.4)

	1982 to 83 Sex and number of subjects	
	Male 105	Female 82
Range	ND† to 1.5	ND to 2.3
Mean	0.10	0.09

† ND = not detected; limit of detection was 0.01 mg/kg.

Several other organochlorine compounds are used by industry and these warrant further investigation as it is now clear that the organochlorines can be very persistent in both the environment and the food chain, unless the chemicals are highly volatile. For example 1,1,1-trichloroethene used in dry cleaning fluid is unlikely to give residues in food unless it is used close to the food chain, because it is so volatile. But there is a need to widen the scope of organochlorine analysis in food from the rather narrow range of substances presently studied, which are primarily pesticides or former pesticides plus other organochlorines such as PCBs that are detected in routine analysis for organochlorine pesticides.

3.4 POLYNUCLEAR AROMATIC HYDROCARBONS (PnAHs)

This is a large group of substances with the common structural feature of two or more fused benzene rings (Fig. 3.4). (They are also known as polycyclic aromatic hydrocarbons because of their cyclic structures.) It is well established that PnAHs are environmental contaminants and that they are produced during combustion. Less well established is whether they are produced in other ways. There have even been claims, from experiments on hydroponic cultures in sealed environments, that they are produced by plants. But this seems unlikely given the energy required to produce the complex chemical structures involved and the lack of any known biological role for them.

PnAHs have been found in a very wide range of foods, and in virtually every country in which they have been studied. It is probable that they are ubiquitous environmental and food contaminants in much the same way as polychlorinated dibenzo-*p*-dioxins (section 3.2). Elevated levels are usually associated with smoked food—indeed very high levels can be achieved by some traditional smoking processes. However, in countries where such foodstuffs do not form a major part of the diet, more general environmental contamination may provide the main source of dietary intake. This assumption was tested in the UK after a study indicated that oils and fats, together with cereals, provided the single greatest input of PnAHs to the average diet. PnAHs were traced back through margarine production to oilseed growth. Refining the oil had little or no effect on the

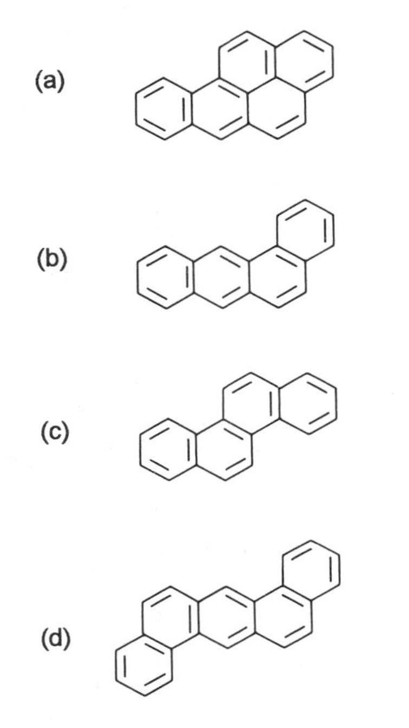

Fig. 3.4. Structures of some polynuclear aromatic hydrocarbons. (a) benzo[*a*]pyrene; (b) benz[*a*]-anthracene; (c) chrysene; (d) dibenz[*a,h*]anthracene.

levels of PnAHs in the finished margarine. The source appeared to be on the farm, leaving environmental or natural sources as the most likely ones.

If one dismisses natural production (see above), the fallout of PnAHs on crops from combustion processes seemed the most likely cause—with probably some preferential transfer into the oil since PnAHs are lipophilic.

PnAHs are comparatively easy to detect in foodstuffs and so a large number of different members of the group have been studied. However, the data on many of the PnAHs are difficult to interpret since they do not appear to be present in any pattern—no particular ones seem to predominate in food samples. Various ways of overcoming this problem have been proposed, including concentrating analytical effort on those PnAHs that are known to be toxic. This is quite sensible since not all of them may present a hazard in food, and there is clear evidence that some PnAHs are carcinogenic. Alternatively, workers have measured only one or two 'marker' PnAHs usually benzo[*e*]pyrene or benzo[*a*]pyrene (Fig. 3.4) but this does not give an adequate measure of risk as some other carcinogenic PnAHs are not included. Unfortunately an approach using toxicity weighting factors, such as that used for polychlorinated dibenzo-*p*-dioxins and dibenzofurans would not be viable for PnAHs as genotoxic carcinogenic effects cannot be weighted. As a result there are now considerable data on PnAHs in food in many countries, but it is not possible to provide a clear picture of the resulting toxic risk to which consumers are exposed. This is very unfortunate given the effort involved in generating all the data and as there is clearly intake at quite high levels of several PnAHs that are carcinogenic. This risk is only partly avoidable since combustion is never likely to be lessened to the extent that food is not contaminated with PnAHs. Although exposure to them cannot be eliminated, work on PnAHs in food appears to be lacking objectives much of the time. The general objective must be the same as for other environmental chemicals in food—to reduce the levels of PnAHs in food as much as possible.

3.5 OTHER ORGANIC ENVIRONMENTAL CHEMICALS

The range of organic environmental chemicals being studied in food is very small compared with the vast number of these substances that might be found in food. As discussed earlier (section 3.1) the problem is one of deciding which of these chemicals to devote analytical resources towards. A number of these other chemicals have been proposed for analysis in food over the last ten years or so, and these are discussed below.

Benzene is now accepted as a carcinogen and, taken together with its widespread use, this probably makes it worth studying in food. But there has been little research on whether it is found in food. Its volatility probably requires studies to concentrate on those parts of the food chain that are close to likely sources of benzene. These include chemical manufacturing facilities and sites where food is produced or sold close to benzene release. It should not be too difficult to develop suitably sensitive methods of analysis and it would probably be worthwhile including some of the other simpler aromatic chemicals such as toluene in this work, particularly where there is suspicion that they are toxic to man.

There has been some evidence that alkyl, aryl and alkyl-aryl phosphates, which were used to replace PCBs (section 3.3.1), may be environmentally persistent and possibly toxic. By analogy with PCBs it may be worth carrying out food surveillance for these substances around industrial waste disposal sites, although it is likely that alkyl phosphates are more readily degraded, by hydrolysis, than their aromatic counterparts. It might be possible to adapt existing methods of analysis for organophosphorus pesticides so that surveillance can be carried out.

Phthalate esters (Fig. 3.5) have most usually been studied in food because of their use in plastic packaging. But they are now more widely used in plastic that is not used in contact with food, for example in furniture. And they are released from plastic into the environment. For example chemical analysis usually detects a considerable background of phthalate esters from the immediate laboratory environment. Although they are now used less in food packaging than before and their usage in other materials probably leads to food contamination, the latter route has yet to be fully explored. Methods of analysis are available from previous work on their migration from packaging and could be applied directly to look for environmental contamination of food. By analogy there are probably several other chemicals in plastic whose levels in food might be explored via the environmental as well as the packaging route.

In addition to these candidate chemicals, there are probably many more which need to be considered if it is accepted that environmental contamination could lead to organic chemicals contaminating the food chain, at levels that might in some cases pose a risk to health or at least in such a way as to stimulate action to reduce such contamination to as low a level as possible (the ALARA [= as low as reasonably achievable] approach). This principle is not yet fully established amongst those involved with studying food chemical contamination, but with interest in environmental protection growing it is likely to be increasingly discussed.

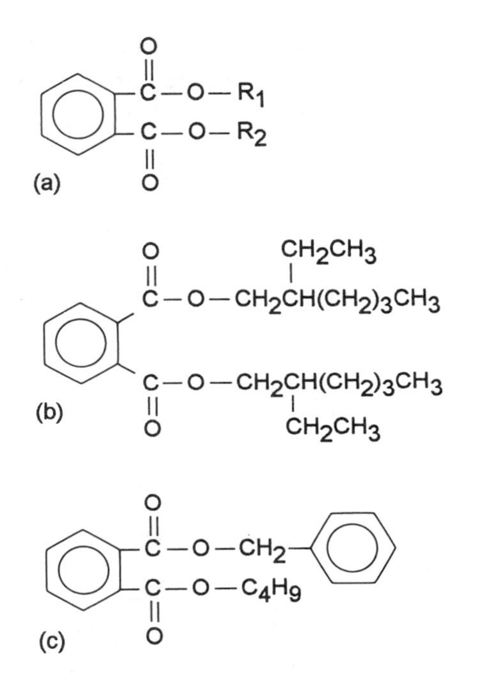

Fig. 3.5. Phthalate esters. The generic structure (a) can have the same substituent groups (R_1, R_2), as in di-2-ethylhexyl phthalate [structure (b)] or different groups, as in butylbenzyl phthalate [structure (c)].

3.6 CONCLUSIONS

There is firm evidence that both industrial chemicals (e.g. PCBs) and by-products (e.g. PCDDs and PCDFs) can contaminate the food chain. And as analytical methodology and environmental monitoring extend to look at more of these chemicals, work on them in food is also likely to develop. However, it is essential that this work is well focused. Without a systematic approach to selecting the chemicals to study, work on organic environmental chemicals in food is likely to stay in established areas of activity—for example looking in greater detail at PCBs in food—although the sources of and exposure to many of these chemicals have been researched extensively. The type of prioritization scheme described in section 3.1 of this chapter needs to be developed and applied. There also needs to be parallel growth of the necessary databases on the toxicity, environmental persistence and other information about the chemicals produced and/or used by industry, and on chemical by-products of industrial activity. The size of the task is enormous and it is probably necessary to reduce it by some pragmatic measure such as limiting the variables involved in judging which chemicals to study, for example by setting a lower limit of 1 tonne of annual production for chemicals to be included in the selection process. This will inevitably exclude some chemicals that contaminate food, but probably not the majority.

Thus at present we know a lot about a few chemicals that contaminate food from the environment. There is clearly more work to be done on some of these, notably PCDDs and PCDFs, before even the general picture about how they contaminate food is complete. There is much to be done before we can even define how we are exposed to them via the food chain. Studies on how to reduce our exposure can only follow such work. Another key lesson we have learnt is that source-directed measures are most likely to be effective in controlling organic environmental chemicals in food. This requires tracing back contamination along the food chain and through the environment to the sources. The chemicals involved are too stable and persistent to control them at a later stage. The main problem confronting investigators is the apparent multiplicity of sources in some cases, notably for PCDDs and PCDFs.

FURTHER READING

Dioxins

Food Surveillance Paper No. 31. HMSO, London. (1992)

Pollution Paper No. 27. HMSO, London. (1989)

CCMS Report No. 176, NATO/Committee on Challenges to Modern Society, Environmental Protection Agency, Washington DC, USA. (1988)

Ahlborg, U. G., Håkansson, H., Waem, F. and Hernberg, A. *Rapport frän en nordisk expert grupp*. Nordiska Miniterrådet, Miljörapport **7**. (1988)

Unweltbundesamt und Bundesgesunheitsamt. *Sachstand Dioxine, Stand November 1984*. E. Schmidt Verlag, Berlin. (1985)

Eadon, G. E., Kaminsky, L., Silkworth, J., Aldous, K., Hilker, D., O'Keefe, P., Smith, R., Gierthy, J., Hawley, J., Kim, N. and DeCrapio, A. *Environmental Health Perspectives*, **70**, 221–7. (1986)

PCBs

Food Surveillance Paper No. 13. HMSO, London. (1983)

HCB, DDT and HCH

Worthing, C. R. (ed.) *The pesticide manual—a world compendium*, 8th edn. The British Crop Protection Council, Thornton Heath, UK. (1987)

Food Surveillance Papers Nos. 16 and 25. HMSO, London. (1986, 1989)

Polynuclear aromatic hydrocarbons

Dennis, M. J., Massey, R. C., McWeeny, D. J., Knowles, M. E. and Watson, D. *Fd. Chem. Toxicol.*, **21**, 569–74. (1983)

Bartle, K. D. Analysis and occurrence of polycyclic aromatic hydrocarbons in food. In: *Food contaminants: sources and surveillance*, RSC, Cambridge, UK. (1991)

4

Nitrate, nitrite and *N*-nitrosamines

D. H. Watson, Ministry of Agriculture, Fisheries and Food, R242, Ergon House, c/o Nobel House, 17 Smith Square, London SW1P 3JR, UK.

4.1 INTRODUCTION

Nitrate and nitrite are of interest not only because they are used to inhibit toxin formation in food by *Clostridium botulinum* (Chapter 5, section 5.2.4) but for a variety of other reasons, not least that their presence in the food chain leads to the formation of nitrosamines, some of which are carcinogenic in experimental animals. There have been extensive chemical studies on the reactions of nitrate and nitrite in food, and there has been considerable analytical work to define the levels and incidences in the diet, and hence consumer intake, of these two ions. Although the main purpose of these studies has been to identify the main sources in the diet of nitrate and nitrite, another line of research has developed on their contribution to the formation of nitrosamines, by nitrosation, before and after the food has been eaten. There is now growing evidence that endogenous nitrosation takes place in humans—this was in doubt for many years—and so it is particularly important that dietary exposure to nitrate and nitrite is fully examined.

There are several sources of dietary nitrate and nitrite. These ions are used as food preservatives because of the inhibition of toxin production by *Clostridium botulinum* by nitrite—added nitrate is itself inactive until it undergoes reduction to nitrite. There is formation of nitrate in crops from the use of nitrogenous fertilizer and from other inputs in the nitrogen cycle. There is also a natural background level of nitrate in the general environment and this probably leads to some nitrate in the diet. As well as these dietary sources, nitrate is present in water and in body fluids. This multiplicity of sources is unusual for a chemical contaminant and makes it essential to carry out extensive studies on the origins and fates of dietary nitrate, and nitrite, if a clear picture of the risk is to be provided.

Nitrosamines are formed by the nitrosation of organic amines, or other organic amino substances such as ureas or amides (Fig. 4.1). The nitroso (-NO) group may react with nitrogen, carbon or sulphur atoms, but it is the *N*-nitroso group (N-NO) which has attracted greatest attention because of the carcinogenicity of several compounds

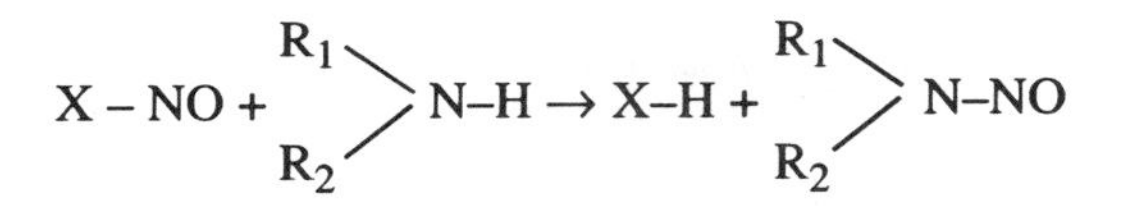

Fig. 4.1. Nitrosation.

containing it. The range of *N*-nitrosamines found in food is very great, but attention has concentrated on *N*-nitrosodimethylamine (NDMA; Table 1.2) which has been shown to induce cancer in a wide variety of laboratory animals, and hence is probably a carcinogenic hazard to man. The other *N*-nitrosamines have been studied less extensively in both toxicological work and food analysis, but they are believed by some to pose similar hazards to NDMA where they are present in food, or in the environment. A list of some of these other *N*-nitrosamines is given in Table 4.1, together with some of the targets for their carcinogenicity. It should be noted that at least one compound, *N*-nitrosoproline, is probably not carcinogenic. Analysis of food for nitrosamines has concentrated on volatile *N*-nitrosamines. Until recently methods were not available for the detection of non-volatile *N*-nitrosamines in food. However, now that these non-volatile compounds can be detected there is growing evidence that they may provide the majority of dietary intake of *N*-nitroso compounds. There has also been recent work on the formation of nitrosamines in the human body. This has helped to develop knowledge about our potential exposure to nitrosamines as a group but the picture is still far from complete.

The study of nitrate, nitrite and nitrosamines in food provides part of a larger picture about human exposure to these substances. Account must be taken of other potential sources including the environment, the work place, and inhalation of nitrosamines from tobacco products. For example, a specific group of alkaloids in tobacco is nitrosated, including nornicotine and anabasine. Their *N*-nitroso derivatives are known to be

Table 4.1. Some *N*-nitrosamines

	Carcinogenic†	Main target organs
N-nitrosodiethylamine	√	Liver, kidney, lung, oesophagus, forestomach
N-nitrosodi-*n*-propylamine	√	Liver, lung, oesophagus
N-nitrosodi-*n*-butylamine		Liver, bladder, forestomach, oesophagus, lung
N-nitrosopyrrolidine	√	Liver, lung
N-nitrosopiperidine	√	Liver, GI tract, respiratory tract
N-nitrosomethylbenzylamine	√	GI tract
N-nitrosodiethanolamine	√	Liver, respiratory tract
N-nitrosoproline	×	—

† After oral exposure in rats and/or mice.

carcinogenic in the lung, nasal cavity, trachea and/or oesophagus. Together with polynuclear aromatic hydrocarbons (Chapter 3, section 3.4) and probably also polychlorodibenzo-*p*-dioxins and dibenzofurans (section 3.2) these carcinogenic *N*-nitrosamines probably account for some of the cancer-inducing effects of cigarette smoking and other forms of tobacco usage. It is the association between smoking or chewing tobacco and cancer that has provided much of the impetus to research on *N*-nitrosamines.

Although nitrate and nitrite are most usually studied in the context of *N*-nitrosamine production, nitrite itself is of interest because at high levels it can cause methaemoglobinaemia—particularly in infants under three months old. This is caused by reaction of nitrite with haemoglobin which makes the blood less effective in transporting oxygen. However, this is a rare condition in the UK and most other countries.

Since the study of nitrate and related compounds covers a wide range of topics, the rest of this chapter concentrates on the three central issues in work on these substances in food: how they enter the food chain, the extents to which they are present in food and possible ways of reducing consumer intakes of *N*-nitrosamines.

4.2 NITRATE AND NITRITE

4.2.1 Dietary sources

The natural uptake and use of nitrogen by living organisms and its use as a food preservative are the 'direct' sources of nitrate in the diet. The use of nitrogenous fertilizers is one contributor to the first of these two sources. Others include fixation of atmospheric nitrogen and the use of manures (Fig. 4.2). The complexity of nitrogen uptake and usage in the biosphere is underestimated by those who claim that the use of artificial fertilizers in agriculture is the sole source of nitrate in food or water. The use of fertilizer undoubtedly contributes but not in a simple way. For example Table 4.2 summarizes the results of an extensive study on the effects of fertilizer application, at different rates, on nitrate concentrations in different crops. In this experiment crops were harvested at intervals after nil, optimum, and one and half times optimum application rates of fertilizer. (The optimum application rate was the minimum rate at which maximum yields were expected.) The results given in Table 4.2 are for nitrate levels in samples harvested at the end of the trial—when nitrate concentrations would have been expected to be closest to equilibrium with the inputs and outputs of the nitrogen cycle. The results show that nitrate concentrations were not necessarily raised by fertilizer application when this was optimal. However, one and a half times the optimal rate of application led to elevated nitrate concentrations in cabbage and beetroot, for which there was also evidence that nitrate levels were raised after optimal application of fertilizer, and in carrots and leeks where optimal rates did not seem to alter nitrate levels.

Many factors can influence the levels of nitrate in crops. These include the following:

- The season in which crops are grown. Experiments comparing nitrate concentrations in crops grown in winter under artificial lighting seem to indicate that nitrate levels in crops are higher under such conditions.
- Increased irrigation may decrease nitrate concentrations in crops.

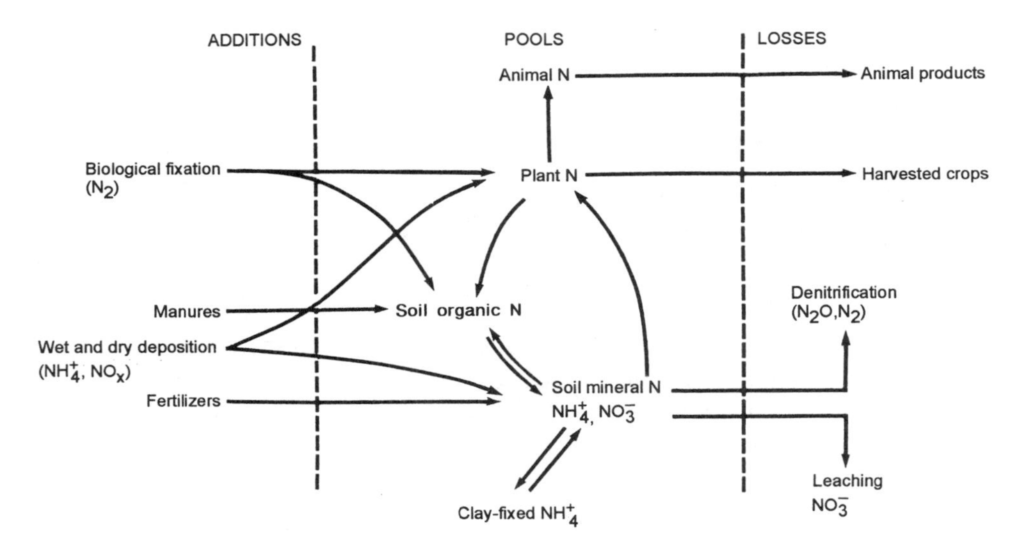

Fig. 4.2. The nitrogen cycle.

— Storage of vegetables has unpredictable effects. It can increase or decrease nitrate concentrations, or it may not alter them. No clear pattern has emerged from experiments in several countries.

— Cooking may reduce levels of nitrite in vegetables but it is not clear what the effects of cooking are, if any, on the more chemically stable nitrate ion, although there are some indications that cooking might reduce its levels too.

— Lengthening the period between nitrogen fertilizer usage and crops being harvested might be expected to decrease nitrate levels in crops, as fertilizer nitrogen works its way around and out of the nitrogen cycle (Fig. 4.2). But this is not a general rule.

Table 4.2. Effect of fertilizer application rates on nitrate concentrations in various crops

Crop	Nitrate concentrations (mg/kg) in crop with:		
	No fertilizer	Optimum application rate	1.5 times optimum rate
Cabbage	120	420	1300
Carrots	2	2	52
Leek: shank	4	3	34
flag	8	9	2
Onions	96	67	80
Beetroot	15	590	1800
Potato	42	19	42

What appears to happen is that nitrate levels stay low in an initial period, after fertilizer application, and then rise, plateau and decline quickly. Experimental work supporting part of this thesis is summarized in Table 4.3, where a decline in nitrate levels was found. The data in this figure derive from experiments involving the application of optimal rates of fertilizer, as defined above. Where 1.5 times optimal rates were used the picture was similar, although as one might expect nitrate concentrations declined more slowly when more fertilizer was used (Table 4.4).

— The effects of organic farming on nitrate levels in crops are not established. A direct comparison of the effects of using manure or chemical fertilizer on nitrate levels in crops, of the same variety grown under otherwise similar conditions, has not apparently been carried out. There are nevertheless claims that organic farming leads to lower nitrate levels in vegetables, although there are very wide variations in nitrate levels even in controlled experiments.

Table 4.3. Effect of period between fertilizer usage and harvesting on nitrate levels (mg/kg) in crops, using optimal application of fertilizer

Cabbage		Leek (shank)		Beetroot	
Date	Nitrate	Date	Nitrate	Date	Nitrate
12/6/87	4900	11/10/87	100	31/7/87	2200
22/6/87	4600	21/1/88	45	12/8/87	1200
2/7/87	3900	22/3/88	3	21/8/87	970
13/7/87	2900			3/9/87	1000
23/7/87	1600			14/9/87	590
13/8/87	420				

Table 4.4. Effect of period between fertilizer usage and harvesting on nitrate levels (mg/kg) using 1.5 times optimal fertilizer application rate

Cabbage		Leek (shank)		Beetroot	
Date	Nitrate	Date	Nitrate	Date	Nitrate
12/6/87	5100	11/10/87	200	31/7/87	2600
22/6/87	5200	21/1/88	130	12/8/87	1600
2/7/87	6000	22/3/88	34	21/8/87	1500
13/7/87	5500			3/9/87	900
23/7/87	4200			14/9/87	1800
13/8/87	1300				

Since there are many variables influencing nitrate levels in crops, it is not surprising how widely nitrate concentrations can vary from even sample to sample of the same variety of a crop. And levels vary markedly between different vegetables. The statistical distribution is therefore unlikely to be a normal Gaussian one, where one would expect the median and mean to be the same. The median and mean concentrations of nitrate in some surveys differ. A skewed, non-Gaussian distribution would be consistent with the many variables involved. Studies why this may be the case for nitrate could provide the first detailed examination of an asymmetric distribution of contaminant levels in food surveys (see Chapter 11, section 11.3).

The other known source of dietary nitrate and nitrite is their use in food preservation. The active agent is nitrite and there has been a clear trend towards its usage rather than using nitrate (from which nitrite is formed in food by microbiological and biochemical processes). However, nitrate/nitrite mixtures are still used in some commercial 'curing mixes'. As well as inhibiting the formation of botulinum toxin, nitrite imparts colour and flavour to cured meats. Its use stems from the traditional use of saltpetre (potassium nitrate) to cure meat several centuries ago. There are European Community controls on the maximum permissible levels of nitrate and nitrite that can be added to cured meats, including bacon and ham, and these have been enacted in the UK via a series of *Preservatives in Food Regulations*. A recent survey showed that one out of 213 samples of bacon, ham and other cured meats exceeded the statutory levels (Table 4.5). The sample was a ham product. When these results for retail samples were compared with data from a survey of the amounts of nitrate and nitrite present in cured meats at the factory, the levels of nitrite in retail samples were generally less than those in samples from the factory, although in other respects the results of the two surveys were broadly comparable. This could be taken as supporting the general contention that, if anything, nitrite levels in cured meats decline following curing. But more work is probably needed to examine this further. Although this should be reassuring for those concerned with exposure to nitrite, it could have important implications for inhibiting botulinum toxin formation since there is a general trend towards decreasing nitrite levels as far as possible in

Table 4.5. Survey of retail samples of cured meats for nitrate and nitrite

Product	Nos. of samples	Nitrate (mg/kg)	Nitrite (mg/kg)
Bacon	51	ND–310†	ND–76
Ham	59	3–410	ND–110
Other pork products	49	ND–59	ND–130
Beef products	23	4–30	2–15
Liver products	6	7–26	ND–11
Chicken products	3	6–22	ND–11
Turkey products	4	13–31	1–84
Tongue	18	ND–82	1–71

† ND = not detected; limits of detection were 0.8 mg/kg (nitrate) and 0.5 mg/kg (nitrite).

cured meats, and any further drop in nitrite levels after products leave the factory might possibly lead to the preservative action being lost.

Nitrate is also used as a preservative in the production of some cheeses (Table 4.6). It is used to stop taint from bacterial coliforms and anaerobic bacteria during ripening. The resulting nitrate levels in cheese are low (usually less than 50 mg/kg; compare concentrations in some cured meats: Table 4.5). This is perhaps because nitrate is only necessary as an aid in producing the cheese and is not intended to help preservation of the cheese once it has ripened.

Table 4.6. Types of cheese in whose production nitrate usage is recognized in international standards

Danablu	Esrom
Danbo	Herrgärdsost
Edam	Hushällsost
Gouda	Norvegia
Havarti	Maribo
Samsoe	Fynbo
Tilsiter	Amsterdam
Limburger	Leidse (Leyden)
Saint-Paulin	Friese
Svecia	

In addition to the above 'direct' inputs to dietary nitrate and nitrite, there are 'indirect' sources, such as those leading to nitrate in water. This is important since water is of course used in many areas of food production, from brewing to diluting fruit concentrates. Nitrate in beer derives from water, and hops and other ingredients. Hops are probably the main source of nitrate in beer. Given efforts to reduce nitrosamine levels in beer, several breweries in the UK have probably reduced the nitrate contents of brewing ingredients.

4.2.2 Dietary intake

Dietary intakes of nitrate have been estimated in several countries (Table 4.7). Similar results are given by different studies and it is likely that the average intakes of nitrate in these countries are well within the acceptable daily intake (ADI) of 0–219 mg/person per day for a 60 kg person (derived from the European Community (EC) Commission's Scientific Committee for Food's (SCF) ADI of 0–5 mg/kg bodyweight for nitrate ion). It is possible to extrapolate from the dietary studies on which these estimates are based to derive estimated intakes for groups with specific dietary habits. For example this has been done for various types of vegetarians, which is important given the major contribution of vegetables to dietary intakes of nitrate in some countries. The UK estimates for nitrate intakes by vegetarians are given in Table 4.8. Again these estimated intakes are within the ADI but it is noticeable how similar they are and how they are between three

Table 4.7. Estimates of average dietary intake of nitrate

Country	Estimated average intake from the diet (mg/person per day)†
UK	52–53
USA	73
Switzerland	72
Germany	49

† Excluding drinking water and non-alcoholic beverages: the UK estimate if these are included is 54–61 mg/day.

Table 4.8. Estimates of dietary intakes of nitrate by vegetarians in the UK

	Estimated intake (mg/person per day)
Vegans	185
Demi-vegetarians	191
Lacto-ovo-vegetarians	194

and four times the estimated average intake for the general population (compare Tables 4.7 and 4.8). This underlines the central finding of dietary studies on nitrate, that vegetables can provide a consistently large proportion of dietary intake. This is probably not the case in some countries where cured meats may provide a major part of dietary intake, but where such meats form a small part of the average diet, as in the UK, vegetables provide about three quarters of the average dietary intake of nitrate.

Dietary intake of nitrite is generally much less than that of nitrate. In the UK average intake is in the order of 2 to 5 mg/person per day. Unlike nitrate, it is common not to detect nitrite in the majority of food samples in a general dietary survey. And where cured meats are not widely consumed, vegetables could be expected to be the single greatest dietary source of nitrite. Intake is close to but probably within the EC Commission's SCF's ADI for nitrite of 0–0.07 mg/kg bodyweight per day.

Although most studies of dietary exposure to nitrate and nitrite depend on the analysis of food, it has been proposed that urinary nitrate excretion might be used to estimate nitrate intake. But to be able to extrapolate from such work to study dietary intake, 'residual' nitrate in urine needs to be considered. This nitrate is found in the urine of volunteers eating 'nitrate-free' diets. Whilst it is debatable whether any food can ever be free of a ubiquitous ion such as nitrate, 'residual' nitrate in urine needs to be taken into account if urinary nitrate excretion is to be related to intake. If this is done there seems to be a reasonable correlation between average nitrate intake from the diet at about 50 to 75 mg/person per day (Table 4.7) and urinary excretion at about 88 mg per person per day, if the latter figure is corrected for the presence of 'residual' nitrate in urine of about 12 mg/person per day. The variations in such figures from person to person probably need to be established fully if the technique is to be widely accepted. Nevertheless studies on urinary nitrate levels for subjects with achlorhydric (neutral pH) stomachs—who excrete less nitrate—tend to support this type of approach, although they indicate that comparison of nitrate intake is not possible for patients with this condition.

Dietary intake of nitrate cannot be readily compared with intakes from other sources. Although such sources would appear to be minor, on the basis of urinary nitrate excretion, there are clearly intakes from drinking water and probably also via inhaled nitrogen oxides. Intake from drinking water probably varies between different countries and between regions of the same country. And whilst human intakes of nitrogen oxides from the air can be estimated, it is not known to what extent these oxides are converted in the body to nitrate or nitrite. However, it is clear that smoking and other forms of pollution can contribute significantly to the inhalation of nitrogen oxides.

Thus a considerable amount is known about food sources of nitrate and nitrite and the relative exposure of some major population groups, particularly vegetarians, compared to the general population. This provides largely reassuring information. Less is known about other sources of exposure to nitrate and nitrite, although on present evidence the diet appears to be the biggest contributor to nitrate and nitrite intakes by man.

4.3 *N*-NITROSAMINES

4.3.1 Dietary sources

Although there has been considerable research on nitrosamines, it is not possible to provide a comprehensive picture of the factors involved in their formation in food. Much is known about the chemistry of their formation but this work has generally been carried out *in vitro* rather than on food.

The nitrosation reaction (Fig. 4.1) requires a nitrosating agent to be formed from nitrite. The nitrous acidium ion (H_2O^+NO), dinitrogen trioxide (N_2O_3) and dinitrogen tetroxide (N_2O_4) act as nitrosating agents. Their potency in nitrosation varies with pH. *N*-nitrosation of secondary amines has been studied in greatest depth. Its kinetics are defined by the following equation: $\text{Rate} = K_3[R_2NH]\,[\text{nitrite}]^2$. The reaction is highly pH related. It depends, for example, on the ionization of reactive groups in the molecule to be nitrosated. The kinetics of the initial *N*-nitrosation of primary amines are similar. Tertiary amines are nitrosatable but the rate and course of the reaction depend considerably on the amine's structure. The nitrosation of biological molecules, for example polypeptides and bile acid conjugates, has been demonstrated but not extensively studied. Given the dietary importance of polypeptides, for example, it would be valuable to carry out further research on the nitrosation of such molecules.

The exact reasons for the presence of different *N*-nitrosamines in food are not entirely clear. Where there have been detailed studies on food or beverage raw materials some interesting information has been produced. Two notable examples are studies on *N*-nitrosamine formation in cured meats and alcoholic beverages.

— *Cured meats* *N*-nitrosopyrrolidine and *N*-nitrosodimethylamine are the *N*-nitrosamines most consistently found in fried bacon. Research has shown that these two compounds are found mainly in the fat of fried bacon. But this is probably not the result of higher temperatures in the fat than in the meat of bacon rashers. (Frying at 185°C gives highest levels.) It might be due to the formation of intermediate pseudonitrites of unsaturated lipids. Reducing frying temperature decreases production of these two nitrosamines, with no evidence of formation at 100°C.

— *Alcoholic beverages* The presence of *N*-nitrosamines in whisky and beer has been extensively studied. The use of natural gas (instead of coal gas) has been identified as a major factor in their formation. Malt from kilns fired with natural gas can have high levels of *N*-nitrosodimethylamine (NDMA). This is because nitrogen oxides are formed in the very high temperature regions of the natural gas-air flame. Temperatures in the region of more than 1500°C are required. Modification of natural gas burners to produce 'cooler' flames led to lower nitrogen oxide levels and consequently major reductions in NDMA levels in malt. Alternatively burning sulphur in the malting process can be used to inhibit NDMA formation. The low sulphur content of natural gas helped to allow NDMA formation when the use of higher sulphur content coal gas was replaced with natural gas in the UK. The precursors of NDMA and other nitrosamines in malt have not been identified but the tertiary amines hordenine and gramine (Fig. 4.3) are implicated.

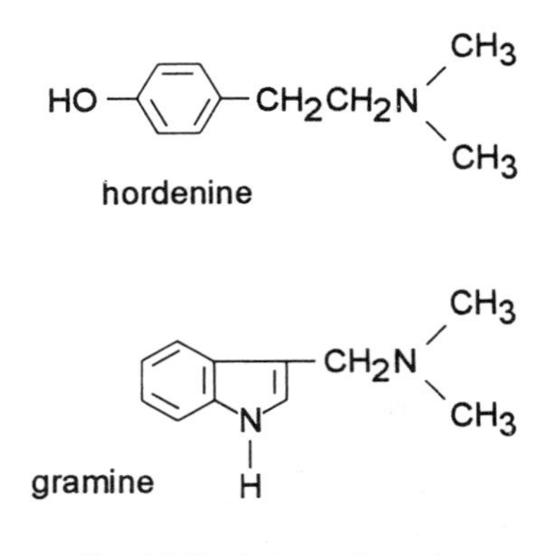

Fig. 4.3. Hordenine and gramine.

These studies illustrate the value of applying an empirical approach to identifying the causes of *N*-nitrosamine formation in food or drink. This does not necessarily require a detailed understanding of the reaction kinetics or mechanisms which, for the nitrosation of chemicals in complex mixtures in food or drink, are likely to be very complicated and extremely difficult to reproduce *in vitro*. Hence much of the free solution chemistry that has been done to explore *N*-nitrosamine formation may turn out to be of marginal help in defining the production of these substances in the diet. A pragmatic, empirical approach using food or drink samples is more likely to provide useful information as in the above work on cured meats and alcoholic beverages.

An empirical approach has also proven its worth in identifying those dietary components in which volatile *N*-nitrosamines are found. They make up a relatively small group of foods and beverages.

— *Cured meats* Volatile *N*-nitrosamines have been found at levels generally in the μg/kg range (parts per 10^9). Analytical effort has concentrated on the volatile *N*-nitrosamines, particularly *N*-nitrosodimethylamine and *N*-nitrosopyrrolidine. These have been found, *inter alia*, in bacon, and in canned, cooked and uncooked cured meats. Other *N*-nitrosamines detected in cured meats include *N*-nitrososarcosine, *N*-nitrosoproline, *N*-nitrosohydroxyproline and *N*-nitrosothiazolidine-4-carboxylic acid.
— *Fish and fish products* *N*-Nitrosamines have been detected in relatively few samples. *N*-nitrosodimethylamine has been detected in untreated and cured fish. The low incidence of detection is surprising given the variety of organic amines that have been found in fish and fish products.
— *Alcoholic beverages* *N*-Nitrosodimethylamine (NDMA) has been widely detected in beer and those spirits derived from malted barley (for the reasons noted above). The levels are, again, usually in the μg/kg range. Surveys of UK beer over the past decade or so indicate a decline in levels of NDMA as a result of the action described above, with levels now close to or less than 1 μg/kg.
— *Fermented foods* NDMA has been found at low μg/kg levels in some cheese samples. Cheese, yoghurt and other cheese fermentation products might be expected to contain *N*-nitrosamines particularly where nitrite or nitrate is added during fermentation (section 4.2.1).

Given the ubiquitous nature of nitrate in the environment, it is surprising that volatile *N*-nitrosamines seem to be found in such a restricted range of foods and beverages, and that where they are detected the levels are several orders of magnitude less than those of nitrate and organic amines in food. This has led to the development of methods of analysis for 'apparent total *N*-nitroso compounds' (ATNC) in food. These methods provide data on the concentrations of the N-NO moiety in food and beverages. Comparison of such data with those for individual *N*-nitrosamines in food (Table 4.9) shows that there may well be many more *N*-nitroso compounds in food than those which are currently identifiable one by one.

Analysis for ATNC in foods and beverages has become more widely accepted over the last few years, but this has not provided a radically new perspective on *N*-nitroso compounds in the diet. Although ATNC appear to be present in a range of foods and beverages, their levels are higher by a large margin, in some cases, than those of volatile *N*-nitrosamines or the few non-volatile *N*-nitrosamines that can be detected (Table 4.9). A wide range of ATNC levels have been found in beer (for example up to 570 μg (N-NO)/kg with a mean of 54 μg (N-NO)/kg). ATNC levels in UK beer have decreased, in the same way as volatile *N*-nitrosamine concentrations have declined (see above). However, there is no obvious relationship between their levels and nitrate concentrations in beer. In contrast ATNC concentrations in US-cured meats appear to have increased. Studies by the US Food and Drug Administration indicate that this may be because some US manufacturers have not been maintaining a high degree of quality control in bacon curing.

Table 4.9. Mean levels (μg/kg) of apparent total *N*-nitroso compounds (ATNC) and individual *N*-nitroso compounds in some samples of various foods and beverages

Samples	ATNC	*N*-Nitrosoamino and imino acids	Volatile *N*-Nitrosamines
Fried bacon	1400	108	11
Cured meat	910	29	0.2
Beer	150	10	3
Tea	7	Not detected (detection limits: 1 to 20 μg/kg depending on substance being analysed)	Not detected (detection limit: 0.1 to 1 μg/kg)
Coffee	20		
Drinking chocolate & cocoa	42		
Dried soup	14		
Dried milk	8		

4.3.2 Dietary intake

It is debatable whether estimated intakes of *N*-nitroso compounds are of help in studying the toxicology of these substances. Estimated intakes cannot take account of each *N*-nitroso compound in the diet since analytical methods are not available for many non-volatile *N*-nitrosamines. And information about the toxicology of *N*-nitroso compounds is restricted to relatively few compounds; for carcinogenic *N*-nitroso compounds no observed effect levels have not been defined (and may not be definable). Nevertheless estimated intakes can provide a useful picture of how intake is changing over time and point to those foods or beverages which provide major inputs to intake of these compounds.

In the UK average intakes have been estimated as follows:

— For NDMA an intake of 0.1 to 0.6 μg/person per day has been estimated, with alcoholic beverages providing the biggest contribution to this intake.

— A figure of 36 μg (N-NO) per person per day has been estimated for the average intake of ATNC by adults. The 97.5 percentile intake was estimated as 140 μg. Beer was proposed as the largest contributor. These intakes were about an order of magnitude above those for volatile *N*-nitrosamines.

4.3.3 Inhibiting the formation of *N*-nitrosamines

Given the potential risk of cancer from these compounds, it is not surprising that there has been considerable work to try and find ways of preventing their formation. Successful work on reducing NDMA levels in alcoholic beverages is described above (section 4.3.1). There has also been work on chemical inhibitors, notably vitamin C (ascorbic acid). This vitamin is a potent inhibitor of *N*-nitrosation and there are several epidemiological studies which suggest that its intake is inversely correlated with the risk of gastric cancer.

Ascorbic acid has, however, been shown to have little or no effect on gastric nitrosation, perhaps because it is only present in the stomach for a short time following ingestion. Further experimental work is needed to find a way of exploiting the inhibitory effect of vitamin C on nitrosation.

Another route may be to reduce the conversion of nitrogenous material to nitrate in crops. If nitrate levels can be reduced in vegetables, the major dietary contributor to nitrate in several countries, this might indirectly reduce our exposure to *N*-nitroso compounds. But the recent introduction of nitrification inhibitors in agriculture, to prevent nitrite formation in soil by *Nitrosomonas* bacteria and hence reduce environmental nitrate levels, may not significantly reduce nitrate levels in plants. Their use could have other consequences since very little is known about uptake of the inhibitors—nitrapyrin and dicyandiamide—by crops, or their toxicology. It is important not to use substances of unknown toxic potential to try and reduce exposure to substances of known toxicity—this could add to, rather than reduce, risk.

4.4 CONCLUSIONS

The chemistry of nitrate, nitrite and *N*-nitroso compounds, particularly *N*-nitrosamines, has been studied in detail. However, their behaviour and fate in the food chain are, surprisingly, little understood compared with many other chemical contaminants. Nevertheless there is considerable information about the levels and incidences of nitrate and nitrite in the diet and from this intakes can be estimated. These estimates indicate that our exposure to them in food is probably safe. This cannot be said for *N*-nitroso compounds in the diet, partly because of evidence that this class of substances contains carcinogens and partly since it is very difficult to be sure that all of the substances of interest have been identified. This is very similar to the state of research on many other chemical contaminants in food, with the clear exception of 'man-made' contaminants, for example pesticides and veterinary drugs, which usually have quantifiable intakes and extensively researched toxicology.

FURTHER READING

Food Surveillance Papers Nos. 20, 24 and 32. HMSO, London. (1987, 1988 and 1992 respectively)

Pollution Paper No . 26. HMSO, London. (1986)

The health effects of nitrate, nitrite and N*-nitroso compounds*. National Academy of Sciences, 1981. National Academy Press, Washington DC, USA. (1981)

Hill, M. J. (ed.) *Nitrosamines—toxicology and microbiology*. Ellis Horwood. (1988)

Bartsch, H., O'Neill, I. and Schulte-Hermann, R. (eds) *Relevance of* N*-nitroso compounds to human cancer: exposures and mechanisms*. IARC Scientific Publication No. 84. IARC, Lyon. (1987)

5

Toxicants occurring naturally in food

D. H. Watson, Ministry of Agriculture, Fisheries and Food, R242, Ergon House, c/o Nobel House, 17 Smith Square, London SW1P 3JR, UK.

5.1 INTRODUCTION

The terms 'naturally occurring toxicant' or 'toxin' are used to describe a wide range of chemicals of natural origin. In this context 'toxicant' and 'toxin' are used to mean a substance with toxic effect. Toxin is usually used for such chemicals produced by bacteria, algae and fungi. Whilst toxicant is sometimes also used for these substances it is more usually applied to toxic chemicals produced by vascular (higher) plants. In this chapter both terms are used interchangeably since there is no difference in their general significance.

The study of naturally occurring toxicants is directly relevant to chemical contamination of food. Toxic hazard cannot be ignored because the contamination is not the result of human activity. Indeed some naturally occurring toxicants are present in the food supply because they are produced by organisms—algae, bacteria and fungi—that contaminate the food chain and should be controllable by man. Nevertheless there is a school of thought that treats naturally occurring toxicants as a separate group of substances to which well established methods of studying chemical contamination of food cannot be applied. This is recognized but not accepted here since the systematic approach used in studying other environmental contaminants, such as dioxins, or agrochemical residues such as pesticides residues, can provide a firm basis for studying naturally occurring toxicants in food. Indeed, a systematic approach is needed in this research as the natural toxins are a diverse and often too little studied group of substances.

Some believe that natural toxins may pose a greater health risk to the consumer than many, if not all, man-made chemical contaminants in food. This contention is not testable at present, because there is too little quantitative information about the toxicology and intakes of natural toxicants from food. Only by further well targeted research will the importance or otherwise of these chemicals be established. It is with the hope that a more systematic and fruitful approach will develop, that this chapter considers current information about naturally occurring toxicants. The chapter summarizes and reviews what is currently known about this potentially important group of chemical contaminants in food.

5.2 THE DIFFERENT TYPES OF NATURAL TOXICANTS

5.2.1 Introduction

Natural toxicants in food are usually classified by the types of organisms that produce them. The main groups are:

— higher plant toxicants;
— algal toxins;
— bacterial toxins;
— fungal toxins (mycotoxins).

As well as these recognized groupings, there are a few toxicants that are of unknown origin. A well established example is scombrotoxin which is associated with a relatively mild form of food poisoning caused by poor storage of scombroid fish (notably mackerel and tuna). Whilst there have been several hypotheses about the origin of this toxin, including bacterial degradation of these fish, its actual origin has not been identified. In a negative way this underlines the weakness of classifying natural toxicants by their sources. The main problem in studying these substances is in generally having to characterize the toxicant chemically before being able to trace its origin.

The above groupings mainly relate to plants and micro-organisms. There has been some speculation as to why there are apparently no natural toxicants generated by food-producing animals. Certainly several of the natural toxicants produced by plants are parts of their natural defence mechanisms. Their possible effects after we eat plant products may be secondary in biological importance to their natural role in combating pests. Direct evidence for this defensive role comes from observations that several higher plant toxicants are produced in response to stress, for example during drought. In this context the toxicants are referred to as 'stress metabolites' or phytoalexins. However, animals' biochemistry also changes when stress is imposed. And yet none of the 'stress metabolites' produced by animals, for example adrenalin, is widely recognized as a natural toxicant although undoubtedly such substances are present in food. This may be a matter of perspective as much as science. Nevertheless this chapter is limited to the classically accepted groups of natural toxicants in food listed above.

The recognized groupings of natural toxicants in food vary considerably in their chemistry, origins, occurrence in food and effects on man. No common patterns have emerged yet in any of these aspects and this may be the reason why doubt remains about their significance. It also means that the different groups of natural toxicants need to be considered in turn.

5.2.2 Higher plant toxicants

Table 5.1 illustrates the variety of natural toxicants produced by food plants, or by crops that are eaten by food-producing animals. These toxicants can reach the human diet by both direct and indirect routes—from plant products we eat or via animals respectively. Relatively little is known about what happens to higher plant toxicants when they are consumed by food-producing animals, but by analogy with the fate of mycotoxin aflatoxin B_1 (section 5.2.5), they are likely to undergo some chemical changes. One would expect to find structurally related compounds, of possibly different toxicity to the parent

Table 5.1. Some different types of higher plant toxicants

Toxicants	Dietary sources	Example
Pyrrolizidine alkaloids	Several dozen plant species, few of which are consumed as food *per se*, but some are consumed by food-producing animals.	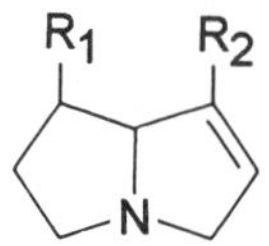 the pyrrolizidine alkaloid nucleus
Psoralens	Wide variety of plants notably the *Umbelliferae* (e.g. celery), legumes and the *Rutaceae* (citrus fruits).	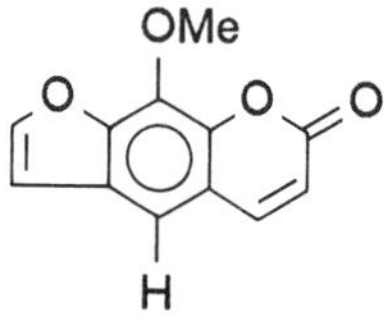 xanthotoxin
'Bracken carcinogen'	Bracken, and possibly some foods produced from animals that have eaten bracken.	Chemicals involved have not been identified.
Glycoalkaloids	Potatoes and, to a lesser extent tomatoes.	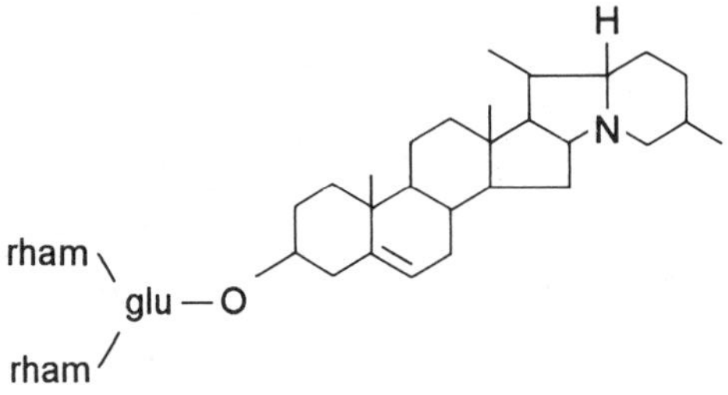 α-chaconine where rham = rhamnose; glu = glucose
Glucosinolates	Brassicas, and hence a wide variety of vegetables, e.g. cabbage, cauliflower, as well as other sources such as rapeseed used in producing animal feed and vegetable oil.	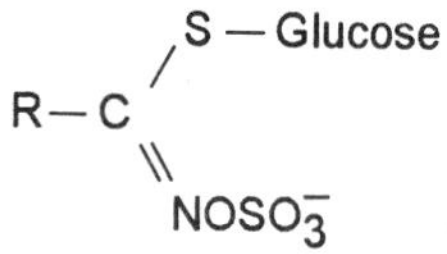 glucosinolate generic structure

toxin, in the meat, offal and milk of food-producing animals if this parallel applies. But at present there is very little information about what actually happens to higher plant toxicants when they are ingested by farm animals.

Table 5.1 also illustrates the variety of chemical structures involved. Although higher plant toxicants are generally complex organic chemicals, there is no other common factor in their structural chemistry. It is difficult to study them because their chemical properties vary so widely. Each time a new set of natural toxicants is identified from a plant source, their chemistry must be explored in considerable detail. This resource-hungry process can drain effort from other important research, notably defining the toxicities of the toxicants. Indeed even where the chemistry of these substances is well advanced, as in the case of all the toxicants in Table 5.1 except 'bracken carcinogen', it is surprising how little is known about their toxicology. So in most cases the risks from our consuming higher plant toxicants are usually unclear and often impossible to quantify on the basis of current information.

The higher plant toxicants in Table 5.1 are considered by some to be of greatest importance for consumers in western Europe. In some other parts of the world the following plant toxicants may be more important:

— *Cycasin* This is produced by cycads which are a source of food starch in some parts of the tropics. The chemistry of cycasin is well explored. There is evidence for it being both neurotoxic and carcinogenic (see *Handbook of naturally occurring food toxicants*. Details of this and other useful further reading are given at the end of this chapter).
— *Oestrogens* A wide variety of higher plants has been shown to contain oestrogenic toxicants. Two groups of plant oestrogens are known: isoflavones and coumestans. Both of these groups have very weak oestrogenic activity but they may be present in quite large amounts in plants. The net biological effect of these substances in food has yet to be defined, but on present evidence it appears to be negligible compared to the influence of the more potent oestrogens produced endogenously by humans.
— *Flavonoids* This large group of substances occurs widely in the plant kingdom, and is most usually associated with providing colour in flowers. However some flavonoids, notably quercetin, are believed to be mutagenic and hence are putative carcinogens. There is evidence that some of these mutagenic toxicants may be found in food, which is not surprising given their wide distribution.
— *The carcinogenic 'principle' in betel nut* A high incidence of oral cancer in central and southeast Asia is thought to be related to chewing betel quid (made primarily from betel nut, betel leaf and lime). However, the chemical compounds involved have not been identified. As tobacco is often used when betal quid is chewed this may be the cause of the oral cancer, at least in part.
— *Possible carcinogens from the* ***Euphorbiaceae*** *and the* ***Thymelaeaceae*** There is limited evidence that some of the members of the two plant families may produce carcinogenic chemicals. For example a high incidence of oesophageal cancer on the West Indian island of Curacao has been associated with a beverage made from the leaves of *Croton flavens*.
— *Toxicant(s) causing favism, lathyrism and coturnism* These three syndromes are associated with population groups in particular parts of the world. Favism is a type of

acute haemolysis which is linked with the consumption of fava beans by genetically susceptible individuals. It has been primarily found in people from countries surrounding the Mediterranean Sea or in the Middle East. The defect in their metabolism has been identified (a deficiency in the enzyme NADP-dependent glucose-6-phosphate dehydrogenase in their erythrocytes). But the toxicants involved are not known, although a group of pyrimidine derivatives is implicated. Lathyrism, a form of muscular paralysis, is associated with eating legumes of the genus *Lathyrus*. It occurs in India and neighbouring countries and may be caused by one or more of a group of unusual amino acids present in the legume. Coturnism, quail poisoning, is probably the least researched of these syndromes. It appears to be limited to a few sensitive individuals in some Mediterranean countries, and has similarities in its effects with favism.

It is interesting that in several of these illnesses the toxicants have not been identified, whereas the examples of importance in Western Europe (e.g. those in Table 5.1) came to prominence from the chemical identification of potentially toxic chemicals in food plants. This difference might possibly be explained by the more limited diets in some parts of the world. Where few different foods are consumed it is possible to identify specific food-related illnesses.

5.2.3 Algal toxins

Algal toxins enter the food chain via other marine species. Food becomes contaminated in the following circumstances:

- when filter feeding shellfish, such as mussels, ingest poisonous algae; or
- when bony fish feed directly on algae or eat other fish that have consumed algae.

Shellfish can become contaminated with toxins from several types of toxic algae:

- *Gonyaulax* species which produce paralytic shellfish toxins.
- *Dinophysis* species which synthesize diarrhoeic shellfish toxins.
- *Gymnodinium breve* which has also been associated with poisonous shellfish with neurotoxic effects.

In each of these types of poisoning the alga is a dinoflagellate whose toxin persists in shellfish after the source organism is dead. The toxins involved are generally very complex molecules with side chains differentiating between the individual substances (Table 5.2). Although their structural chemistry has been studied in some detail, the toxins are generally assayed by biological methods. The most widely used method is intraperitoneal injection of acidified or methanol or acetone extracts of shellfish into mice. The time between injection and the onset of symptoms at a given dilution of extract is used to define the number of 'mouse units' and hence as a measure of the toxicity of the shellfish extract. Work continues on chemical methods of assaying these toxins, but the differing toxicities of members of each group and problems in separating the substances make it very difficult to apply this approach routinely.

Algal toxins are also found from time to time in certain bony fish from tropical and sub-tropical areas. Several hundred species of fish have been implicated, for example moray eel and barracuda which prey on herbivorous fish (e.g. the surgeon fish and parrot

Dinoflagellate	Group of toxins	Chemicals	General structures
Gonyaulax species	Paralytic shellfish toxins	saxitoxin, gonyautoxins, and related compounds	where: R_1 = H or OH; R_2 = H or OSO_3^-; R_3 = A, $CONH_2$ or $CONHSO_3^-$ 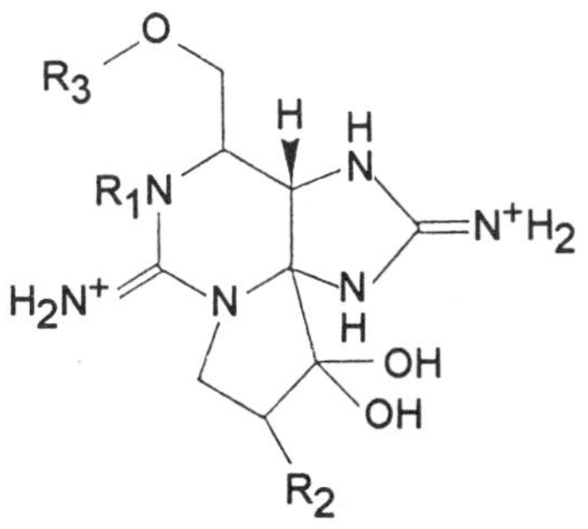
Dinophysis species	Diarrhoeic shellfish toxins	okadaic acid, dinophysistoxins, pectenotoxins	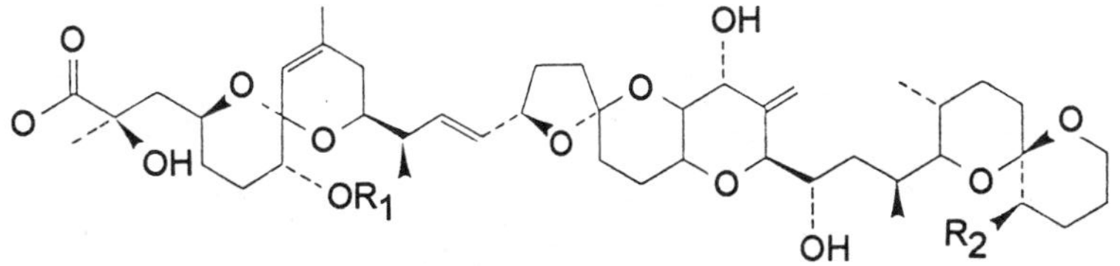 where: $R_1 = R_2$ = H (okadaic acid) where: R_1 = H or acyl; R_2 = CH_3 (dinophysistoxins-1 or -3) 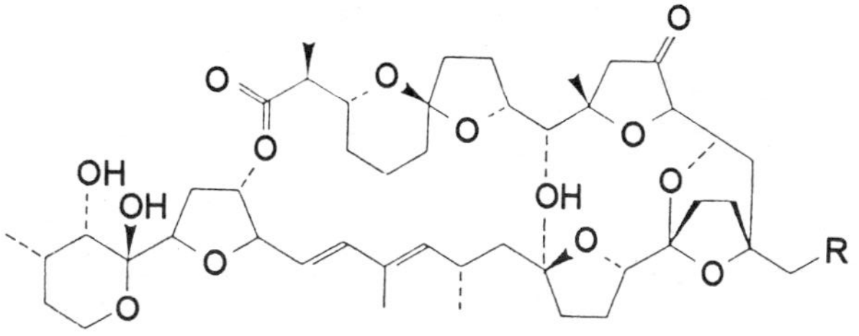 Pectenotoxin-1 (R = OH) or pectenotoxin-2 (R = H)

Table 5.2. (continued)

Dinoflagellate	Group of toxins	Chemicals	General structures
Gymnodinium breve	Neurotoxic shellfish toxins	brevetoxins, GB-3 toxin	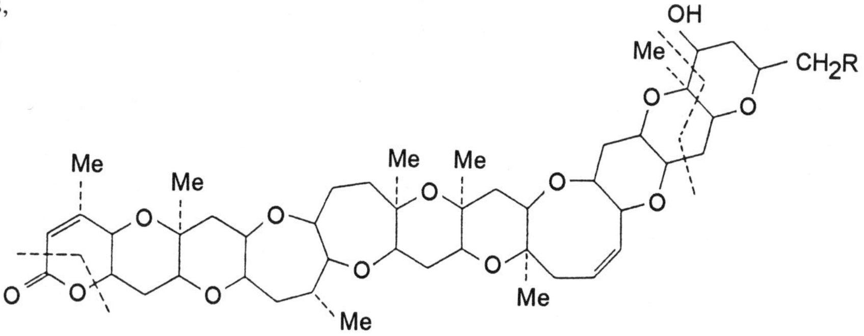 brevetoxin B R = 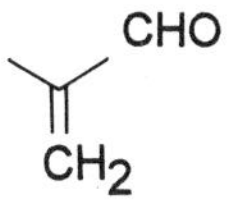 GB-3 toxin R = 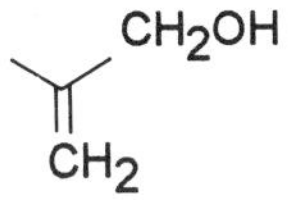 Brevetoxin C R = 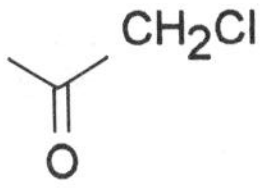

fish) which themselves ingest toxin-producing algae of the dinoflagellate *Gambierdiscus toxicus*. The toxin involved is ciguatera toxin (or ciguatoxin) which may represent a single substance or a family of structurally related toxins. The structure (or structures) is not known. A mouse bioassay is also used to quantify levels of this toxin in fish extract.

All of these algal toxins produce gastrointestinal effects but, with the exception of diarrhoeic shellfish toxins, there are often neurological effects as well. It is not known if there are any algal toxins with more long-term effects but there appears to have been little work on this. In these respects these toxins are analogous to those produced by bacteria (section 5.2.4) and probably distinct from those from higher plants and fungi where chronic effects are believed to be more likely.

Algal toxins are the main group of toxins known to be produced by marine organisms. The other marine toxin that has been identified is tetrodotoxin which is the causative agent of pufferfish poisoning. The meat of pufferfish is considered a delicacy in Japan and China. Preparation of the fish must exclude organs such as the liver which classically contain high concentrations of tetrodotoxin. Levels of this toxin vary from species to species of pufferfish. The toxin has been chemically characterized, is of established structure, and is assayed by a modification of the mouse bioassay noted above, or chemically. There is some limited evidence that it might be of algal origin—pufferfish raised in artificial ponds are free of the toxin—but most authors reject this hypothesis.

5.2.4 Bacterial toxins

These substances are probably the most extensively studied of natural toxicants in food. They are produced by a variety of bacterial species, but the bacteria of greatest importance as regards toxins in food are *Staphylococcus aureus*, *Bacillus cereus* and *Clostridium botulinum*. Other bacteria involved in food poisoning, such as *Salmonella*, do not produce toxins in food. There have been claims that some other bacterial species—such as *Clostridium perfringens*—which are involved in significant numbers of cases of food poisoning produce toxins in food, but this has not been confirmed. Their toxin production occurs after the food has been eaten, and is primarily in the small intestine, or illness is caused by infection rather by intoxication as in the case of *Salmonella* species.

Staphylococcus aureus toxins are of at least seven types—A, B , C_1, C_2, C_3, D and E—with a further less characterized, and hence perhaps not distinct, toxin having been given the letter F. The toxins are all extracellular proteins of similar molecular weights and are resistant to digestion by most proteolytic enzymes—hence their stability in the gastrointestinal tract. Toxins A and D are probably the most common causes of food poisoning by *S.aureus*. However, the presence of toxin is insufficient to identify and trace the bacterial strain causing food poisoning. This has to be done by identifying the bacteriophage(s) (viruses which infect bacteria) associated with the bacterium. The commonest cause of food contamination with *S.aureus* is unhygienic handling; the nose and skin are the usual reservoirs of this organism. The usual symptoms of food poisoning caused by *S.aureus* are vomiting, diarrhoea and abdominal pain which last between twelve and twenty-four hours.

Bacillus cereus has been implicated in two forms of food poisoning. The 'diarrhoeal syndrome' is caused by the production of toxin after the food has been consumed. In contrast the 'emetic syndrome', which generally occurs more quickly, results from ingestion of a comparatively small molecular weight toxin (mol. wt. *c*. 5000) in food, for example rice and noodles. This emetic toxin is very stable. For example it is resistant to digestion by proteolytic enzymes. Its stability and a rapid onset of symptoms (within five hours) are consistent with its causing food poisoning as a result of its formation in food, rather than in the gastrointestinal tract. Its presence in food is thought to result from the storage of cooked food at too high a temperature. This allows any spores of *B.cereus* that are left after the food has been cooked, to multiply and hence to produce enough toxin to cause illness. Much less is known about the chemistry and mechanism of action of *B.cereus* emetic toxin than for the toxins produced by *S.aureus* or *Cl.botulinum*. And at present it is not readily detected in food because assay methods are limited.

Botulism is rare but can be lethal. *Clostridium botulinum* toxins cause this form of neuromuscular paralysis. This usually involves the consumption of only very small amounts of the toxins from foods that have contained spores of the bacterium. The toxins occur naturally in very large complexes of toxic and non-toxic proteins. The neurotoxins themselves are large proteins (mol. wt. $140–170 \times 10^3$). There are at least eight different types—A, B, C_1, C_2, D, E, F and G—of similar amino acid compositions. Toxins, A, B and E have been most commonly associated with botulism. The toxins are composed of 'heavy' and 'light' subunits linked by at least one disulphide bridge. Separating the subunits usually eliminates toxicity. The toxins inhibit release of acetylcholine at the neuromuscular junction hence paralysing control over muscle contraction. The details of how this happens have been studied but the exact mechanism is unknown. This is particularly important as the mortality rate for botulism cannot necessarily be reduced with confidence whilst the fundamental mechanism involved cannot be countered. The best means of control at present rests on preventing toxin production in food. For example nitrite inhibits the production of botulinum toxins and is used for this purpose as a preservative in a range of foodstuffs, notably cured meats (Chapter 4). Ascorbic and erythorbic acids are also known to inhibit toxin production but the minimum effective concentrations of these preservatives are not widely established. But since nitrite is involved in the production of carcinogenic *N*-nitrosodimethylamine and other nitrosamines that might be carcinogenic, ascorbic and erythorbic acids would be preferable provided they are themselves safe and if they could be used effectively to avoid botulism.

The toxins produced by *Clostridium botulinum* and *Staphylococcus aureus* have been very extensively studied compared with most other natural toxins in food. Although their mechanisms of action are still being studied, their other properties have been looked at in considerable detail. And there are several different methods of detecting them. They can be assayed by biological methods, for example mouse lethality for botulinum toxins, and by *in vitro* tests including immunoassay (Table 9.1). Given these advances and the rapid action of the toxins, it is possible to carry out quite extensive detective work to identify the cause of poisoning. However, this is of course limited by other factors such as disposal of contaminated food before the onset of symptoms. The long time courses between toxin ingestion and illness for many of the other natural toxicants in food make it considerably more difficult to investigate their origins and effects on man, even where there are several methods of detecting the natural toxicants involved. For example, it is clearly not possible to test suspect foodstuffs when illness results many years after ingestion of a chemical contaminant, as might be the case for a carcinogen such as aflatoxin (section 5.2.5). For these reasons it can be very difficult to develop methods of preventing or reducing human exposure to natural toxicants in food that are chronically toxic.

5.2.5. Fungal toxicants (mycotoxins)

Over 400 mycotoxins have been identified. The toxins are usually discovered by growing fungi in laboratory culture, testing extracts of them for toxicity using simple assays and then fractionating the extracts and purifying the toxic chemical or chemicals. In most cases the chemistry of mycotoxins isolated in this way is well defined—the structures of most of the 400 plus known mycotoxins have been established. Much less is known, however, about whether many of them are toxic to man and other mammals. About a quarter of the known mycotoxins are believed to have some toxicity to mammals (Table 5.3).

Many fungal genera are implicated but only five, *Alternaria, Aspergillus, Claviceps, Fusarium* and *Penicillium*, have been tested to any great extent for the production of mycotoxins that are toxic to man. Thus the other fungi listed in Table 5.3 may produce other such toxins but considerable resources would be needed to see if this is the case.

The main impetus behind the study of mycotoxins is the search for them in food and raw materials used in food production. For example there has been extensive work on most of the mycotoxins found in food (Table 5.3), and much less effort on the many other mycotoxins—even where there is evidence that they are probably toxic to man. Where a mycotoxin is found in food and is known to be toxic, work can be extensive. Thus much is known about the occurrence in food and animal feed of aflatoxin B_1 and its metabolite aflatoxin M_1 (Fig. 5.1), because it has been well established for many years that aflatoxin B_1 causes liver cancer in mammals. However, much less is known about the toxicology of aflatoxin M_1. Surveys of food for aflatoxin B_1, and the other aflatoxins (B_2, G_1, and G_2) mainly produced by *Aspergillus*, have been carried out in many countries. And there has been extensive research on how aflatoxins are produced in the field, and hence how to stop their contamination of crops. However, many factors are involved in aflatoxin contamination—drought, stress and insect infestation of crops for example—and at present there is no one way of stopping this at source, when *Aspergillus* grows on crops in the field. An alternative strategy of destroying aflatoxins in harvested crops, has also met with limited success. Physical processes such as heating are inadequate, and although chemical treatment using ammonia has been shown to degrade aflatoxins it is not known whether the substances so produced are any less toxic than aflatoxins themselves. At present the only sure way of protecting the food and feed supplies is to reject raw materials containing aflatoxins. This approach can require highly sophisticated technology, with the use of chemical analytical methods that must be capable of detecting parts per 10^9 of aflatoxins in human food or animal feed. This and other types of surveillance for aflatoxins should involve extensive sampling of raw materials since these mycotoxins, and probably others, can be present in only a small part of a batch of food or feed. This is because fungi producing aflatoxins can grow in part of a consignment of food or feed to produce very local, but high levels of contamination. This has implications for the surveillance of food and feed for aflatoxins and other chemical contaminants (Chapter 11, section 11.3.1).

Table 5.3. Sources of mycotoxins that are known to be or may be toxic to mammals

The table lists the fungi known to produce mycotoxins that are toxic in some way to mammals. The metabolites in italics have been found in food. Acutely toxic metabolites from higher fungi (notably toadstools) are not included.

Fungi	Toxic metabolites
Alternaria	*alternariol and related compounds*, *tenuazonic acid*
Aspergillus	*aflatoxins,* aflatrem, ascladiol, asp-haemolysin, aversin, chrysophanol (= chrysophanate), cytochalasins, *cyclopiazonic acid*, emodin, fumitremorgens, gliotoxin, malformin C, *naphtho-γ-pyrones*, β-nitropropanoic acid, *oxalic acid*, paspalanine, physcion and related metabolites, secalonic acid D, *sterigmatocystin and related metabolites*, terreic acid, territrems, TR-1 and TR-2 toxins, tryptoquivaline, tryptoquivalone, viomellein, xanthoascin, xanthocillins, xanthomegnin
Cephalosporium	ophiobolins
Cercospora	unidentified toxins
Chaetomium	chetomin, chrysophanol, cytochalasins, mollicellins C and E
Claviceps	ergot alkaloids, paspalinine
Cochliobolus	ophiobolins
Curvularia	cytochalasins
Diplodia	diplodiol
Dreschlera	chrysophanol, ophiobolins
Engleromyces	cytocholasins
Fusarium	fusarenon-X, *moniliformin*, sporofusarines, *trichothecenes*, *zearalenone*
Helminthosporium	cytochalasins, ophiobolins
Hormiscium	cytochalasins
Metarrhizium	cytochalasins
Micronectriella	*trichothecenes*
Microsporium	viomellein, xanthomegnin
Myrothecium	satratoxins, verrucarin A
Nigrosabalum	cytochalasins
Penicillium	brefeldin, chrysophanol, citreoviridin, *citrinin*, *cyclopiazonic acid*, cytochalasins, funiculosin (= islandicin), hadacidin, janthitrems, *ochratoxins*, *patulin*, paxilline, *penicillic acid*, *penitrems* (= tremortins), PR toxin, *roquefortine*, rubratoxin B, secalonic acid D, simatoxin, viridicatum toxin, viomellein, xanthocillins, xanthomegnin
Phoma	brefeldin, cytochalasins
Phomopsis	cytochalasins
Pithomyces	sporidesmins

Table 5.3. (continued)

Fungi	Toxic metabolites
Rhizoctonia	slaframine
Rhizopus	unidentified toxins
Rosellinia	cytochalasins
Scopulariopsis	unidentified toxins
Stachybotrys	satratoxins, verrucarin A
Trichophyton	viomellein, xanthomegnin
Trichothecium	*trichothecenes*
Verticimonosporium	satratoxins, verrucarin A
Zygosporium	cytochalasins

Some, perhaps all, of the lessons learnt in controlling aflatoxins will probably apply in dealing with the many other mycotoxins that might be found in food. Study of mycotoxins in food, to date mainly aflatoxins, ochratoxin A, sterigmatocystin and the trichothecenes, has provided some probably general points about the production, stability and toxicity of mycotoxins in the food chain:

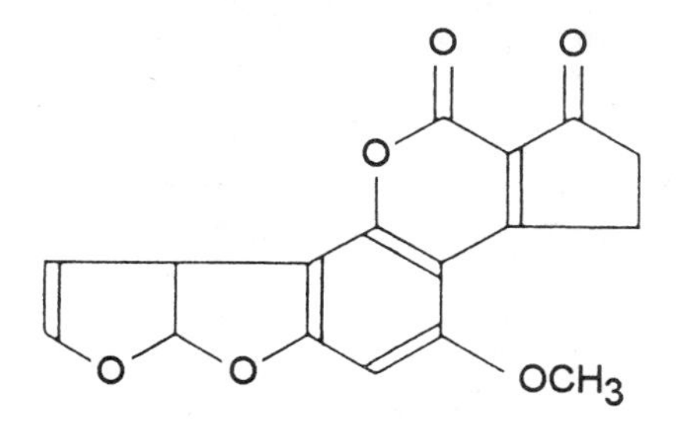

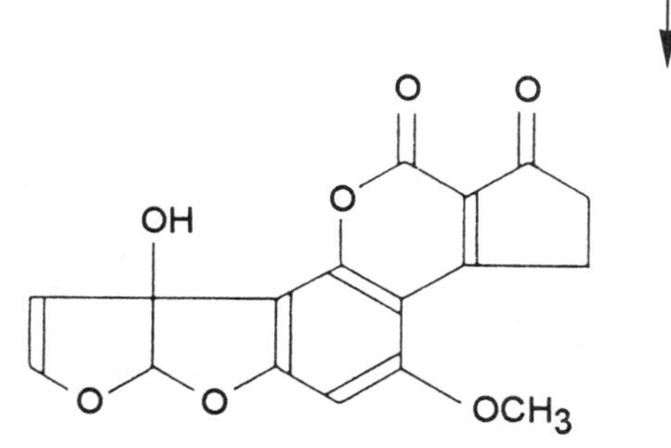

Fig. 5.1. Aflatoxins B_1 and M_1.

— The production of mycotoxins is an unpredictable process. A given mycotoxin is not always produced by one strain of a given fungal species (e.g. an *Aspergillus flavus* strain does not always produce aflatoxin B_1).

— The levels and incidences of mycotoxins in food cannot be predicted with confidence. Food and food raw materials must be tested. This is probably because many factors come into play in mycotoxin production and only some of these have been identified as limiting, for example temperature and water content generally have to be above minimum levels for mycotoxin production.

— Mycotoxins are produced by secondary metabolic pathways—they are not part of the biochemical processes that are essential for life. The reasons why mycotoxins are produced are unclear. However, much is known about the chemical pathways by which a few of them are formed—aflatoxins, sterigmatocystin and the trichothecenes in particular.

— Mycotoxins are relatively stable in the food chain. Even metabolism by animals seems to have relatively little effect on the cores of aflatoxin and ochratoxin molecules, although there are modifications to the chemical side chains.

— The toxic effects of mycotoxins on man cannot be predicted without extensive *in vivo* studies. The range of possible toxic effects of the mycotoxins listed in Table 5.3 is extensive with different mycotoxins damaging internal organs, disrupting blood cells, or inducing tremors, for example. Since the effects are mainly chronic rather than acute, it is very unlikely that a mycotoxin in food would be definitively identified as the cause of a human illness. However, there can be very strong circumstantial evidence where dietary contamination is extensive, as in the case of aflatoxin contamination of food in some tropical and sub-tropical countries where liver cancer is also common. Even so this type of circumstantial evidence needs to be viewed with caution—the high incidence of liver cancer in some parts of the world could be caused in part at least by exposure to type B hepatitis virus.

5.3 HUMAN EXPOSURE TO NATURAL TOXICANTS IN FOOD

The amounts of higher plant toxicants and mycotoxins that we might consume must be estimated from the incidences and levels of contamination, via the analysis of food samples. This is largely done by using chemical methods of analysis or via immunoassays. Human exposure to bacterial and algal toxins is usually measured by the incidences of the respective types of food poisoning, and their presence is still primarily confirmed by biological tests, the results of which can be difficult to quantify. So there is much less information about the levels of these latter toxins in food. It could be argued that there is little need to estimate how much of a bacterial or algal toxin we consume since their effects are primarily quick and well described, but if prevention is to be effective one needs to have quantitative information about current exposure to a natural toxicant in food, if only to allow the effects of preventative action to be monitored. If exposure is not quantified one cannot be sure whether it is reduced when preventative action is taken. Fig. 5.2 illustrates the point—aflatoxin M_1 contamination of the UK milk supply was reduced dramatically after the source of the problem, the supply of groundnut, and its products, containing aflatoxin B_1 was halted.

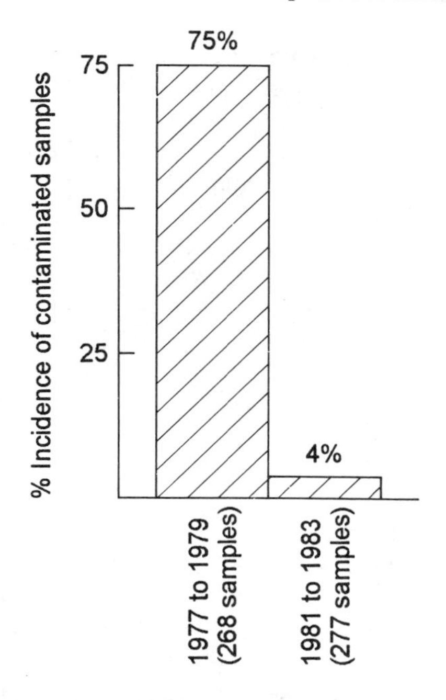

Fig. 5.2. Decreased exposure to aflatoxin M_1 in UK milk. In 1982, the importation and sale for use in animal feedingstuffs of groundnut and its derivatives containing aflatoxin B_1 in excess of 50 μg/kg was prohibited in this country. This had a dramatic effect on the incidence of the metabolite aflatoxin M_1 (Fig. 5.1) in UK milk. The analytical detection limit was *c.* 0.1 μg/kg.

Some mycotoxins and to a lesser extent a very few higher plant toxicants have been subjected to quantitative surveys of relevant foodstuffs. There are extensive data on aflatoxin B_1 in food in many countries, and a growing database on ochratoxin A. Most of the surveillance of food for mycotoxins has been carried out in developed countries. An example of UK surveillance for one of the trichothecenes—deoxynivalenol (vomitoxin)—is shown in Table 5.4. (The structure of this mycotoxin is given in Table 1.2.) The results in this table show the type of information that can be obtained from surveillance of a natural toxin in food and feed raw materials. It is possible to compare contamination in samples from different supplies of a given raw material (e.g. wheat and maize from different countries in Table 5.4) or of various crops (e.g. soya beans, maize and wheat in this table). Provided sufficient samples are analysed and the sampling provides representative material, a picture can be built up and some clues provided about the main source of contamination.

Surveillance for *mycotoxins* and, to a lesser extent, for *higher plant toxicants* has provided the following conclusions about our exposure to the more extensively studied substances in food:

— Aflatoxin B_1 and related aflatoxins can be found in groundnuts, maize, figs and, probably less frequently, in a variety of other nuts or in cereal grain. Aflatoxin

Table 5.4. Survey of deoxynivalenol in imported cereals and pulses (the analytical detection limit was 20 μg/kg)

Sample	Period covered	Numbers of samples analysed	Numbers of samples in range (μg/kg)			
			Not detected	20 to 100	100 to 500	> 500
US soya beans	1982	20	20	0	0	0
Brewer's maize	1980 to 81	10	3	5	2	0
N. American maize	1980 to 83	33	0	2	17	14
S. African maize	1981 to 82	10	3	7	0	0
French maize	1981	4	0	4	0	0
Italian maize	1982	2	0	0	2	0
Canadian soft wheat	1981 to 82	5	0	1	3	1
US hard wheat	1981 to 82	14	2	7	4	1
EC wheat	1982	14	8	4	1	1

contamination is usually associated with crops from hotter climates and in particular where crop rotation is not systematically carried out.

— Ochratoxin A contamination appears to occur primarily in temperate countries and perhaps mainly affects cereal grain. However, there has been much less surveillance for ochratoxin A in food and food raw materials than for aflatoxins. But, like aflatoxin B_1, ochratoxin A is known to be retained in animal products following its contamination of animal feed. Indeed there have been several surveys of pig kidney contamination with ochratoxin A in different western European countries. Surveillance for aflatoxins in animal products has concentrated on aflatoxin M_1 in milk, although quite a lot is known about the other aflatoxins found in the organs of farm animals.

— Trichothecene mycotoxins have been quite extensively surveyed in wheat and several other cereal crops in the northern Hemisphere and southern Africa, with particular emphasis on deoxynivalenol following the discovery of extensive contamination of Canadian wheat with this mycotoxin in the early 1980s. However, subsequent toxicological studies showed that deoxynivalenol is relatively innocuous. This illustrates the need for sound toxicological studies before the decision is taken to survey food for a mycotoxin.

— The above mycotoxins are usually found at between 0.001 and 1 mg/kg in raw materials and in a minority of samples. But there are notable exceptions, for example no deoxynivalenol was found in soya beans in the survey summarized in Table 5.4; aflatoxin B_1 is rarely not detected in groundnut meal.

— Higher plant toxicants, such as glucosinolates and glycoalkaloids, are usually detectable in their known dietary sources (Table 5.1) but it is often not clear whether their ingestion has had any toxic effects on consumers. It seems that exposure to higher plant toxicants is probably greater than consumption of mycotoxins for most people in countries where there is not an endemic problem of fungal growth on crops, but if more extensive data become available on the levels of higher plant toxicants in food this may well prove to be an over-simplification.

5.4 POSSIBLE FUTURE RESEARCH

Research on natural toxicants in food has largely concentrated on their chemistry (Fig. 5.3), particularly their structural and analytical chemistry. There has been less work on their toxicology and comparatively little on ways of reducing consumer exposure to natural toxicants in food. Concentration on the chemistry of natural toxicants has helped the quantification of consumer exposure to them, but without a concerted effort in other areas we cannot know what risk this exposure poses. There have been suspicions voiced that natural toxicants may provide a bigger risk than man-made contaminants such as veterinary drug residues, pesticides residues or environmental chemicals in food. But at present there is not sufficient information about the toxicology of natural toxicants to test this assertion. For example Table 5.5 shows that there are doubts about the toxicology of even quite extensively studied toxicants from higher plants.

In the absence of a balanced package of information about most natural toxicants in food it is difficult to judge the relative importance of work on these as opposed to research on other chemical contaminants in food. Consumer perception seems to be that 'man-made' contaminants are a greater concern, but what is the objective answer? Toxicological work on many natural toxicants is needed. It is not enough to identify possible toxic effects, as in the case of many, but not all, mycotoxins, or to have qualitative information about the possible effects of higher plant toxicants. Quantitative surveillance data of the type available for a few mycotoxins and some higher plant toxicants needs to be matched by quantitative toxicological information such as no observed effect levels (NOELs) and tolerable daily intakes (TDIs). It is nowadays essential to establish these parameters for pesticides and veterinary drugs if they are to be licensed, and it should also be required for natural toxicants before their presence in food is quantified. Unfortunately it is usually the presence of natural toxicants in food which stimulates toxicological work. And the toxicology all too often is not sufficiently extensive to provide NOELs

Table 5.5. Toxic effects of some higher plant toxicants

Psoralens	Mutagenic (but may not be mutagenic without light).
'Bracken carcinogen'	Cancer (but there is doubt about whether there is risk other than from direct consumption of bracken).
Glycoalkaloids	Gastrointestinal disturbance
Glucosinolates	Disruption of thyroid function, suspected as possible cause of goitre.

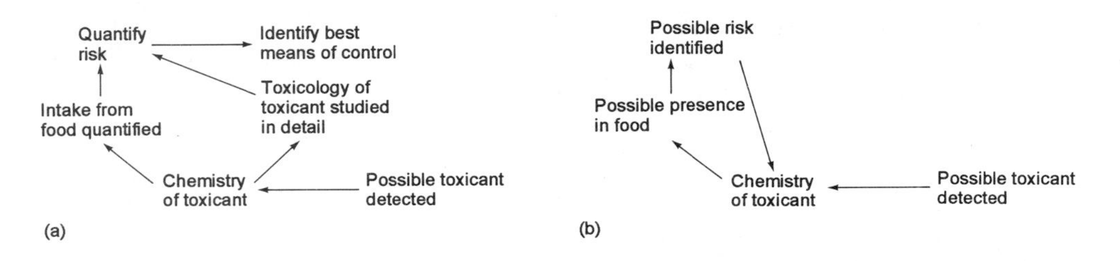

Fig. 5.3. Systematic research on natural toxicants in food. (a) An idealized scheme of research. (b) The approach taken for many natural toxicants.

and hence TDIs. This is presently the case even for many of the more extensively natural toxicants in food—for example for most mycotoxins that are thought to be toxic to mammals (Table 5.6). Unless the toxicant proves to be a genotoxic carcinogen, for which a TDI is usually irrelevant, no real progress can be made without having such quantitative toxicological standards. Critics of this hypothesis would probably note the large numbers of natural toxicants that have been identified and point to the vast resources needed to examine all of them toxicologically. There is no easy answer to this at present. Methods that might reduce the workload, such as studying one member of a large, structurally similar group of natural toxicants or predicting likely toxic effects by looking at the toxicant's structure, are unfortunately not going to help much. There are few distinguishable groups of natural toxicants that have chronic effects, with the notable exception of pyrrolizidine alkaloids, psoralens, glycoalkaloids, aflatoxins and glucosinolates. And even amongst these few groups there is evidence that apparently minor structural changes can have important effects on their toxicities.

Table 5.6. Some claimed toxic effects of mycotoxins (excluded from this analysis are mycotoxins of established toxicity, e.g. aflatoxins B_1, B_2, G_1, and G_2 and ochratoxin A)

Claimed effect	Numbers of mycotoxins for which claim has been made	
Acute toxicity†	15	
Mutagenic	9	No observed effect levels have not been established (applies to Mutagenic through Reproductive organs)
Foetal damage	7	
Anaemia	5	
Internal haemorrhage	10	
Damage to:		
Liver	19	
Kidney	12	
Heart	3	
Spleen	1	
Nervous system	19	
Reproductive organs	2	

† LD_{50} < 100 mg/kg bodyweight on oral administration. (LD_{50} is the dose at which 50% of animals are killed).

One possible way round this problem might be to apply a range of short-term, *in vitro* toxicity tests to a large number of natural toxicants. This could help to reduce the list of candidate substances for more detailed *in vivo* toxicological work. With current *in vitro* toxicity tests this would allow mutagens to be identified. It might also help to identify potent cytotoxins if tissue cultures are used to test natural toxicants. In this way it should be possible to build up a list of candidates for *in vivo* testing. Where there is already substantial evidence that natural toxicants are likely to be carcinogenic or cytotoxic there might be more effort to confirm this via *in vivo* studies. And where there is sound, confirmed evidence that they cause human illness, for example for glycoalkaloids, it is essential to carry out animal studies to quantify the toxic effect and establish a NOEL.

The above proposal does not reduce the need to develop ways of eliminating or at least reducing consumer exposure to natural toxicants of proven toxic hazard. This relatively small group of substances—aflatoxin B_1, ochratoxin A, some of the higher plant toxicants and the algal and bacterial toxins discussed in this chapter—are studied intensively and for some of them, notably aflatoxin B_1 and the algal and bacterial toxins, control mechanisms are used in public health in many countries. But there are continuing needs to improve available methods of control and to develop new ones. The best approach, although rarely the most practicable, is to halt toxin production. Only for *Clostridium botulinum* toxins is this put into practice, for example with the use of nitrite to inhibit toxin production when meats are cured. For the other natural toxicants noted above this is not the favoured approach at present. Mycotoxin production on crops in the field or in storage can be better controlled by screening produce. Part of the reason for this is that the parameters controlling mycotoxin production, or the synthesis of higher plant toxicants, have not been drawn together in such a way as to allow toxin production to be inhibited in a controlled way. This would be a major task. Hence control of natural toxicant contamination is likely to continue to be by established methods of screening and treatment after harvest. This might change if the toxic risks involved prove to be quantifiable and there is sufficient pressure to research further the factors involved in toxin production.

5.5 CONCLUSIONS

Natural toxicants in food are in some cases known to pose a risk to human health. But this is the case for relatively few of the known toxicants and mainly those with fairly immediate effects such as bacterial and algal toxins. The risks in these cases are not readily quantifiable and perhaps there is no need to do this since the case for control is clear—indeed control mechanisms protect the population against these toxicants in many countries. It is more difficult to assess the risks posed by toxicants from fungi and higher plants. There are some clear examples of established toxic hazards for a few of this large group of chemicals but further examination, including quantification of the toxicological effects of these substances, is probably the most sure way of putting them in perspective. It would also help in identifying their relative importance as chemical contaminants in food. The diversity of chemical structures and possible toxic effects for natural toxicants will not make this an easy exercise. Continued research on the few mycotoxins known to be hazardous should continue but it needs to be put in context—there is always the suspicion that at least one or two of the hundreds of less well studied mycotoxins may prove to be more hazardous. Indeed some higher plant toxicants (e.g. those in Table 5.1) probably occur at greater levels in the western diet than most other natural toxicants, and yet there is little point in further refining the estimates of exposure to them without a concerted effort to define what this means in toxicological terms. This is not to say that study should not cease, rather it requires the context to be more meaningful.

For many years scientific work on natural toxicants has been less extensive than that on some other chemical contaminants in food, notably residues of pesticides in food. There now seems to be growing interest following acceptance of the idea that there may be natural or inherent risk in the diet. It is important that the opportunity is not lost and that effort is concentrated on answering the key question—how great is the risk?

FURTHER READING

Watson, D. H. (ed.) *Natural toxicants in food: progress and prospects*. Ellis Horwood. (1987)

Hirono, I. (ed.) *Naturally occurring carcinogens of plant origin—toxicology, pathology and biochemistry*. Elsevier, Oxford, UK. (1987)

Rechcigl Jr., M. (ed.) *Handbook of naturally occurring food toxicants*. CRC Press, Boca Raton, Florida. (1983)

Aquatic (marine and freshwater) biotoxins. Environmental Health Criteria, **37**, WHO, Geneva.

Food Surveillance Paper No. 18. HMSO, London. (1987)

Food Surveillance Paper No. 36. HMSO, London. (1993)

6

Chemicals migrating from food packaging

S. R. Pugh, Ministry of Agriculture, Fisheries and Food, R239A, Ergon House, c/o Nobel House, 17 Smith Square, London, SW1P 3JR, UK.

6.1 INTRODUCTION

With the introduction of plastic packaging in the 1960s came interest in the migration of organic chemicals from the packing into the food. Initial concerns were voiced when dairy products, packaged in polystyrene, contained a distinct taint (organoleptic effect) from styrene monomer which had migrated from plastic into the food. In the 1960s packaging companies and plastics manufacturers were being asked, by Government and consumers, to assess the toxicity of compounds migrating from packaging into food. A further problem was identified when questions about the toxicity of vinyl chloride became prominent in the 1970s. It was discovered that workers involved in operations using large quantities of vinyl chloride were developing a rare liver disease. In 1978 the then newly formed Steering Group on Food Surveillance (Fig. 9.2) produced *Food Surveillance Paper* No. 2 (HMSO, London) on the contamination of food by vinyl chloride monomer. In the same year a Statutory Instrument limited the amount of vinyl chloride monomer that could be found in food contact plastic in the UK to 1 μg/kg of plastic. Styrene migration from polymers made from styrene monomer has also been of interest. Work on these polymers was reported in *Food Surveillance Paper* No. 11 (HMSO, London). This paper detailed the uses of styrene-based polymers in food contact materials and reported on the migration of the monomer into food. The Committee on Toxicity of Chemicals in Food, Consumer Products and the Environment (COT) reviewed the toxicity of styrene in that report. The COT encouraged the efforts being made to reduce levels of styrene monomer in food. Recently the toxicity of styrene has been further reviewed in the EC and calls made to reduce styrene levels still further. As interest and awareness grew the problem of contamination of food with chemicals migrating from packaging began to be actively studied in many countries.

6.2 EC CONTROLS

Although most of the research work in the UK is concentrated on plastic, this is not the only food contact material from which chemicals may migrate into food. However, international controls have tended to concentrate on plastics, notably in the EC. The process of developing legislation on this area of work in the EC is probably as far advanced as anywhere in the world. The EC controls are likely to include several hundred chemicals used in the manufacture of plastic food packaging. The Commission of the EC has produced a series of directives limiting the substances allowed in food contact plastics. Within the EC there is a list of over 90 chemicals that have been assessed as safe for food contact use and within certain criteria, and are allowed in food contact materials. Many other chemicals used in plastic food packaging are currently being reviewed. These controls are based on information about the chemicals used within the EC and their possible toxic effects. The Commission has proposed that, within a few years, all of the chemicals used to manufacture food contact plastics should be controlled by positive lists, that is lists of substances that can be used.

As well as prescribing the maximum amounts of monomers and other starting substances that might migrate from a plastic, the EC directives also give the maximum overall migration from a food contact material (overall migration is the total amount of material that transfers from the packaging material into a food). The EC controls stipulate the use of simulants, rather than foods, so that a plastics manufacturer can relate the results of testing to all foods and conditions of use. The food simulants are said by the EC Commission to emulate the properties of the foods that the packaging is likely to encounter. One of the directives (85/572/EEC) relates migration into food simulants to the migration that might be expected into particular foods. The overall migration test simply limits the total weight of material that migrates from the plastic. Thus it is not really a safety check but more a measure of quality of the plastic. Indeed, the overall migration test simply measures by weight the total amount of material migrating under the conditions of the test. Directive 90/128/EEC puts the limits as 10 mg/dm^2 of packaging or 60 mg/kg of food. The simulants given in the directive are water, 3% acetic acid, 15% ethanol—these simulants mimic water-based or aqueous foods—and olive oil which mimics fatty food.

The overall migration test into aqueous simulants involves leaving the plastic material or article in contact with the simulant for 10 days at 40°C, or as stipulated by the Directive for specific conditions of use (82/711/EEC). The simulant is then evaporated by leaving it in an oven at temperatures above 100°C until the weight of the residue is constant. The weight of the residue after evaporation of the simulant is then converted to give the overall migration figure, in weight of migrant per unit area of plastic (mg/dm^2) or per unit weight of food (mg/kg). Overall migration into olive oil is also measured gravimetrically and is complicated by the absorption of olive oil by the plastic. The experimental procedure is as follows. The polymer sample is weighed and then immersed into olive oil. At the end of a prescribed time the material is removed from the olive oil and any simulant adhering to the surface of the sample is removed. To assess the amount of oil absorbed by the plastic the oil is extracted using pentane. The extract is then hydrolysed, using an alkaline solution, and finally converted into the methyl ester. The amount of olive oil absorbed in the material is then determined by gas chromatographic analysis of the ester solution. This figure is then subtracted from the final weight of the same polymer after it has been exposed to olive oil, giving a revised figure that takes into account the absorbed olive oil. Thus overall migration by this test is the initial weight of the plastic minus the revised figure for the final weight after the migration test. These tests are described in more detail by the Comité European de Normalisation (CEN) in their test methods for measurement of overall migration.

6.3 RESEARCH ON CHEMICAL MIGRATION FROM CONTACT MATERIALS INTO FOOD

Scientific work on the study of chemical contamination of food from contact materials and articles has tended to focus on chemicals that have known toxicological properties, as in the case of styrene and vinyl chloride (section 6.1). Food contact materials are complex substances with diverse natures and most scientific work has concentrated on plastics because they are a less traditional packaging material when compared with paper or glass, for example. However, the distinction between the different packaging types (paper, plastics, glass and metals) is becoming less important as the materials are used in combination. For example many tin cans are fabricated with a resin inner layer, and much food contact paper has a plastic coating.

6.3.1 Plastics

The many ways in which plastic packaging is used means that in addition to its structural/polymeric components there are a number of substances added to the plastic to modify the properties of the polymer. Examples of lower molecular weight substances used in plastic packaging are listed in Fig. 6.1 and include: plasticizers (substance (a)), antioxidants (substance (b)), ultraviolet light stabilizers and absorbers (substance (c)), antistatic agents (substance (d)), impact modifiers, blowing agents (substance (e)), lubricants, slip agents (substance (f)), biocides (substance (g)) and preservatives. Although these chemicals are not all used at the same time, provision should be made for any one of these substances migrating from plastic. The complications involved in measuring all of the substances migrating from a single plastic are illustrated by looking at experiments that attempted to analyse all of the chemicals migrating into food. The number of possible migrants makes this an involved piece of work. One type of polymer for which it was attempted was thermoset polyesters. This polymer is heavily cross-linked by styrene. It is designed for use at high temperatures, both in microwave and conventional ovens. With this type of material 1000–1500 mg of volatile chemical migrants were identified per kilogram of plastic. These volatile substances were those with molecular weights of less than 200. Of the less volatile components, with molecular weights greater than 200, the amount of material migrating from the polymer was between 1000 and 2000 mg/kg of plastic, of which palmitic acid was the largest component. Table 6.1 lists more than 62 of the volatile compounds, the majority of which have been confirmed using mass spectrometry as migrating from thermoset polyesters. These include many of the functional plastics additives (for example numbers 59 and 60 in Table 6.1), as well as their transformation products (substances 30 and 31). There are residues of monomers (substances 7, 21, and 54), cross-linking agents (substances 14, 21, 23, 27 and 32), and oligomeric forms of the polymer (substances 61 and 62). These experiments were carried out at about 150°C with the result that there were also degradation products migrating from the

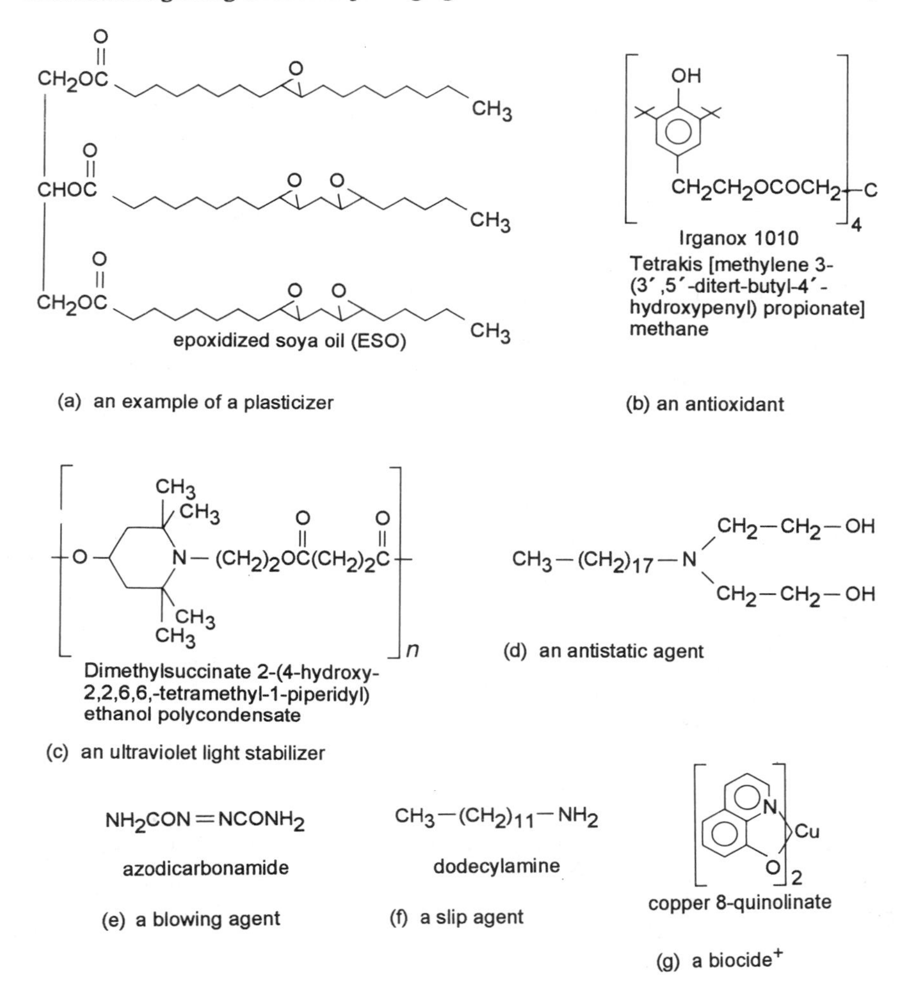

Fig. 6.1. Chemical structures of some common additives used in food contact plastics.

polymer. The high concentrations of oligomers found in the polymer might be due to the breakdown of the polymer. The list of migrants includes benzene, ethylbenzene, benzaldehyde and styrene which are potentially very toxic. The concentrations of these four migrants were measured in a number of thermoset polyesters from different sources. The results are shown in Table 6.2. However in this particular experiment samples of the polymer were heated at 150°C for one hour, an extreme condition that might not be expected in normal use.

Table 6.1. Chemicals identified as migrating from thermoset polyesters

1	1-Hydroxy-propan-2-one (acetol)	33	*tert*-Butoxybenzene
2	Benzene	34	Allyl phenyl ether†
3	Isopropyl acetate	35	Methylbenzoate
4	2-Ethoxyethanol	36	2-Ethylhexanoic acid
5	1- or 2-Heptene (2 isomers)†	37	2,2,4,4,6,8,8-heptamethylnonane†
6	Methyl methacrylate	38	Propiophenone
7	Propyleneglycol	39	Naphthalene
8	Toluene	40	Methyl salicylate
9	2-Ethyl-4-methyl-1,3-dioxolane, (2 isomers)	41	*n*-Butoxybenzene
		42	2-Allylphenol†
10	Dimethyl-1,4-dioxane, (3 isomers)†	43	1,1-Diphenylethene
11	Ethylbenzene	44	Butylated hydroxytoluene (BHT)
12	*m*- and/or *p*-Xylene	45	1,3-Diphenylpropane
13	3-Heptanone	46	Nonylphenol
14	Styrene	47	Diphenylcyclobutane, isomer (a)†
15	*o*-Xylene	48	Diphenylcyclobutane, isomer (b)†
16	3-Heptanol	49	1,2-Diphenylethene
17	Isopropylbenzene	50	1,2-Diphenyl-2-(2-methylpropoxy) ethanone†
18	Benzaldehyde		
19	*n*-Propylbenzene	51	1-Phenylnaphthalene
20	*m*- and *p*-Ethyltoluene	52	2,2-Dimethyl-1,3-propanediol
21	α-Methylstyrene	53	Benzyl alcohol
22	Phenol	54	Propylene carbonate
23	3-Methylstyrene	55	Allylbenzene
24	4-Methylstyrene	56	Eicosane
25	Isobutylbenzene	57	Heneicosane
26	*sec*-Butylbenzene	58	Docosane
27	β-Methylstyrene	59	Hexadecanoic acid
28	Phenylacetaldehyde	60	Octadecanoic acid
29	Dipropylene glycol (3 isomers)	61	1-Phenyl-4-(1-phenylethyl tetralin)(?)†
30	Acetophenone		
31	Styrene oxide	62	1,3,5-Triphenylhex-5-ene(?)†
32	α-Ethylstyrene		

† Compounds awaiting further confirmation by comparison with authentic material.

6.3.2 Paper and board

For these materials the analysis of volatile components is very difficult because of the complex nature of the starting material. The chemical structure of wood is not as well defined as that of many plastics. There are still a number of substances added during processing (additives), such as cleaning agents, coatings and other property modifiers that have yet to be identified. Table 6.3 lists some of the chemicals that have been identified as being present in paper and board. The major components of the volatile migrants

Table 6.2. Levels of specific compounds found by direct headspace analysis of thermoset polyesters

	Concentration of contaminant (mg/kg plastic)			
Sample	Ethylbenzene	Styrene	Benzaldehyde	Benzene
1	24	47	62	0.4
2	6	32	69	0.3
3	14	69	89	0.6

identified in Table 6.3 are alkyl and aryl aldehydes. Their concentrations ranged from less than 10 mg/kg to 35 mg/kg of paper. Compounds identified by solvent extraction were in general present at less than 100 mg/kg. Compounds extracted from paper and board using water, chloroform and ethanol were examined in a screening exercise. A number of the compounds were identified by gas chromatography-mass spectrometry. Many of these compounds were additives and property modifiers. A number of components of printing inks were also identified. Some of these were present even though the paper tested was not printed, suggesting that recycled paper was being used or that there had been some other mechanism for the contaminant being present in the paper.

6.3.3 Metal

There have been changes in can-making technologies over the past few years. Some of these changes were to avoid lead migration from the solder that was used to seal cans, and to avoid tin migrating from the inside layer of a can. These alterations have helped to reduce the general amount of material migrating into the can's contents. But work on testing for migration from cans is made more complex owing to the coatings on the surface. Testing migration from cans using food simulants has been the subject of much research. For example, using one of the simulants, 3% acetic acid, is problematical because even the smallest pinhole in the lining will allow direct contact between metal and acetic acid, causing a reaction and metal dissolves into the simulant. Alternative simulants are being sought, but they need to be validated by comparing them with existing ones.

There is also work on measuring migration from the resins, used to coat cans, which vary from the epoxy resins to polyvinyl chloride co-polymers. It will be very difficult to identify all the possible migrants from these resins. However, it is known that with current tests the coatings probably conform to the overall migration limit for plastic materials. As well as any free monomer migrating from the coating materials there are many other plastics additives in the resin that are likely to migrate. Can-coating formulations include: emulsifiers, corrosion inhibitors, pH modifiers, antioxidants and biocides.

6.3.4 Glass

Work on migration from glass presents different problems from research on the other food contact materials. The majority of the migrating substances are inorganic salts, or

Table 6.3. Chemicals identified as migrating from paper and board

Volatile components	Extracted by solvent
Benzaldehyde	16-Hentracontanone
Benzene	Benzophenone
Butanal	Betulin
Butanedione	
2-Butoxyethanol	
Carbon disulphide	Butylated hydroxytoluene
Chloroform	Docosane
Decane	Docosonol
1,3-Dichloro-2-propanol	Heptadecane
Dihydrodimethyl furanone	Hexadecanol
Ethyl-2,4-dimethyl benzene	
Ethyl acetate	
Furancarboxaldehyde	Michler's Ketone
Heptanal	Octadecanamide
Hexanal	Octadecanoic acid
2-Heptanone	Phenanthrene carboxylic acid
2-Heptenal	
Hexane	
Hexene	
Methyl acetate	Bicyclohexyl phenyl
1-Methoxy-2-propanol	Triphenylmethane
2-Methyl propenal	
2-Pentylfuran	
2-Propylfuran	
2-Methyl-3-pentanone	
Nonanal	
o-Xylene	
Octanal	
p-Xylene	
Pentanal	
Tetramethylpentane	
Trichlorethane	
Trichlorotrifluoroethane	

ions. The major substance leaching from glass was, not surprisingly, silicon dioxide. Table 6.4 lists ions that have been found leaching from glass into water, a simulant defined in EC Directive 82/711/EEC. As the process involved in the reaction between glass and the food or food simulants is probably dissolution rather than migration this is an extreme test of migration. In this experiment the simulant used was stored in the glass containers for 10 days. The simulant was then analysed for migrants using atomic adsorption spectroscopy.

Table 6.4. Ions found migrating from different types of glass into water (Concentrations are in mg/kg.)

	White flint	Amber	Green
Silicon dioxide	12.55	8.84	18.74
Aluminium	0.08	0.07	0.17
Calcium	1.07	0.45	1.76
Magnesium	0.14	0.07	0.11
Sodium	1.57	0.58	0.52
Potassium	0.07	0.05	0.12
Chromium	<0.05	<0.05	0.05
Copper	<0.04	<0.04	<0.04
Iron	<0.07	<0.07	<0.07
Lead	<0.10	<0.10	<0.10
Manganese	<0.04	<0.04	<0.04
Zinc	<0.05	<0.05	<0.05

6.4 TRACE CONTAMINANTS IN PACKAGING MATERIALS

During the manufacture of packaging or the processing of food there is always a potential risk from trace contamination. Some residues of catalysts used in the polymerization reaction during the manufacture of plastic materials could be present in the finished article. The lubricants used in processing machinery in packaging manufacture might find their way into food. However, a great deal of effort is put into removing them from the finished article.

The level of impurities present in a plastic might present particular problems, for example with benzene from the oil feedstock used to make the plastic. The carcinogenicity of benzene makes it essential that it is removed, or at least that its residual levels are reduced as much as possible. The use of benzyl peroxide as an initiator in polymerization reactions did give rise to a potential problem but use of an alternative initiator, *t*-butyl peroxide eliminated this.

6.5 DETECTION OF CHEMICAL CONTAMINANTS IN FOOD FROM PACKAGING

Much of the research on food contaminants from packaging has used headspace gas chromatography, with a variety of detectors for the gas chromatograph. Initially flame ionization and electron capture detectors were used, but with more sophisticated electronics it is now possible to use mass spectrometers and infra-red detectors with an associated increase in the sensitivity of detection. Headspace gas chromatography involves placing a sample in a closed container and allowing the volatile components from the substance of interest to reach equilibrium with the atmosphere in the container. The sample can be heated to increase the rate at which equilibrium is achieved. The vapour above the sample is then injected into a gas chromatography column. This technique works well with volatile contaminants but is clearly less effective for less volatile chemicals, for example styrene. For these less volatile chemicals a modified headspace approach is used. The polymer is dissolved in an organic solvent. Then a 'non-solvent' is added to displace the contaminant from solution. In the case of styrene, water makes an ideal 'non-solvent'. This increases the concentration of contaminant in the headspace vapour which can then be analysed as before. A further variation of the headspace technique is to use a dynamic headspace where the vapour above a sample is continually passed over a sorbant such as Tenax®. Any volatile chemical produced from the packaging is absorbed on the Tenax® which is then placed in a special injection block and the chemicals released through the gas chromatograph.

An effective method for measuring trace organic contaminants is stable isotope dilution gas chromatography-mass spectrometry. This involves the addition of an isotopically labelled analogue of the migratory species to the food, or simulant, in known quantities. The contaminant and analogue are extracted from the food or simulant using, for example, a headspace technique. The resulting extract is then separated by gas chromatography with the contaminant and analogue being eluted as one peak. This peak is then further analysed by mass spectrometry. This is sufficient to separate the analogue from the contaminant. The ratio of the contaminant and analogue peaks from the mass spectrometer is used to give the concentration of the contaminant. The sensitivity of this technique is such that it can be used to determine sub-milligram per kilogram levels of chemical migrants in foods with a high degree of accuracy. Measuring the peak ratios of the contaminant to the analogues using peaks other than the ion peak provides a further safeguard.

The use of food simulants to study migration from materials used in contact with food has increased as legislation to control chemicals migrating into food has developed (section 6.2). The EC Directive 85/572/EEC names four simulants that mimic all types of food: water, aqueous ethanol (15% v/v), acetic acid (3% w/v), and olive oil. Of these simulants the fat simulant, olive oil, is the most contentious. Its relative impurity causes problems when it is used with many of the analytical techniques. There have been other simulants suggested, for example methyl oleate, iso-octane, or Unilever fat HB307 a synthetic mixture of triglycerides. These alternative simulants have been criticized, however. It would be possible if the test material or other factors require it, to use alternative simulants and then validate the alternative simulants against, for example, olive oil. This could be particularly important if a spectroscopic method was routinely being used to test migration from a plastic.

With some chemical migrants there should be no trace of the compound present in the food. Thus the limit of detetection of the most sensitive method of analysis could define the legislative limit on the substance in the food. If the substance cannot be detected at this level it would be deemed not to be present. As limits of detection become smaller then the legislative tolerances would change. For several contaminants limits are already very low. For example analysing for benzene in packaging, a limit of 1 μg/kg can be achieved by isotope dilution and gas chromatography-mass spectrometry techniques using a quadrupole mass spectrometer.

With the increase in accuracy as the analytical techniques became more sophisticated it has become necessary to define a limit of detection (LOD) for some migrants. For lead from cans, for example, a limit of detection of 0.1 mg/kg is quoted in *Food Surveillance Paper* No. 27 (HMSO, London). But in the same report the determinative limit for wet fish was 0.2 mg/kg. The limit of detection is therefore very dependent on the material being analysed, as well as the method used and the degree of accuracy of the analytical results that is required.

6.6 MATHEMATICAL MODELS

The migration of substances from a polymer matrix follows a basic Fickian diffusion model. Thus the rate of diffusion is proportional to the rate of change in concentration of the diffusing species:

$$\frac{\mathrm{d}C}{\mathrm{d}t} \propto \frac{\mathrm{d}^2C}{\mathrm{d}t^2}.$$

Where C = concentration; t = time.

Rearranging this equation to make it more relevant to diffusion of chemicals from polymers, and expressing it as a one dimensional system through the polymer, it becomes:

$$R = -D(C)\frac{\mathrm{d}C}{\mathrm{d}x}.$$

Where: R = flux of migrant (mol/cm^2 s); C = concentration (mol/cm^3); x = thickness of polymer (cm); and $D(C)$ = diffusion coefficient (cm^2/s).

So the rate of diffusion is dependent on not only the change in concentration across the polymer, $\mathrm{d}C/\mathrm{d}x$, but on the diffusion coefficient $D(C)$. The diffusion coefficient is determined by the nature of the polymer. There is a variety of factors that can change the diffusion coefficient for a particular polymer/migration system including the morphology and crystallinity of the polymer, the glass transition temperature and the apparent activation energy of the diffusion process.

Further controlling factors for the degree of migration are:

— the original concentrations of the migrant in the packaging before it was used in contact with food;
— the solubility of the migrant in the contacting phase or the partition coefficient between the polymer and the contacting phase;
— the temperature of the system;
— the contact time.

The properties of the polymer that affect diffusion include such variables as molecular weight, chain branching, density, and affinity for the chemicals that might migrate.

If migration is diffusion controlled, especial care must be taken if, as in the cases of genotoxic compounds, nil migration is required. Given sufficient time a molecule will always have the potential to migrate. Thus if legislation is required to prohibit migration of a particular substance into food, the safe route is not to have the chemical in the packaging.

For the purpose of mathematical modelling, migration from plastics into food has been divided into three classes:

Class 1 —*Non-migration of chemicals* with or without the presence of food.
Class 2 —*Migration* with or without the presence of food (although migration may be accelerated by the presence of food).
Class 3 —*Leaching* which is migration controlled by the presence of food—there is negligible diffusion in the absence of food.

These are not absolute classes to describe chemical movement from plastics but a broad categorization of the types of diffusion likely to occur. One definition of non-migration is where the diffusion coefficient is less than 10^{-12} cm^2/s. Using this approach the diffusion coefficient for leaching would be 10^{-9} cm^2/s in the presence of food and 10^{-12} cm^2/s in the absence of food.

Another approach assumes that the diffusion of chemicals from plastic is a reaction between the migrant and active sites in the polymer. In this case migration involves the sorption and desorption of the migrant from these active sites. This allows the measurement of a rate of reaction and consequently the calculation of a rate constant. Work on this approach concluded that at very low concentrations of vinyl chloride monomer in polyvinyl chloride it was possible to achieve an effective zero level of migration because monomer would be trapped at the active sites. However, the theory did depart from classical Fickian theories in identifying two different types of active sites.

Because the migration of chemical contaminants from polymers is dependent on morphology it is extremely difficult to produce a definitive mathematical model. Attempts to model migration by assuming that diffusion is the underlying process, and incorporating all the morphological changes in the diffusion coefficient, have met with some success. However, they have not helped the legislator to limit diffusion of contaminants into food by incorporating diffusion coefficients into legislation. Legislators are generally limited to using a weight of migrant per unit value for food (mg/kg). This is an absolute weight and gives no relationship to time. To circumvent this problem migration tests are carried out under accelerated conditions to give a figure for migration of the chemical at infinite time, which may be a reasonable approach taking into account the finite amount of time that food is left in packaging.

6.7 ESTIMATION OF INTAKE

Estimating intakes of chemical contaminants in food is a complex topic. For example variations in diet throughout the populations of the world are almost infinite. This subject is described in detail in Chapter 10. For chemicals migrating from packaging it is essential to take into account the extent to which packaging is used as well as the amount of contamination present and the amount of food eaten. Wrapping food in plastics protects the food against many things, for example some forms of microbiological contamination, but there is the risk of chemical contamination from chemicals migrating from the packaging. Their intake must be defined. A broad way of estimating average intake of a chemical migrant from packaging is to take the total production of particular material for food use for a country, and divide this by the number of people in the country. For example, the amount of polyvinyl chloride intended for direct food contact in the UK in 1987 was 13 000 tonnes. As the population of the UK in 1987 was 55 million the average use of PVC was 246 g/person per year. However, this figure is a very broad average and pays no attention to use of the foodwrap; nor does it take account of the consumer who eats above-average amounts of food packaged in PVC, or who uses more packaging materials in the home.

There is a growing amount of information about the amounts of different foodstuffs that are consumed—for example in the UK there is The National Food Survey (a food-purchase-based survey) and the surveys of British adults and other groups, by the Office of Population Census and Statistics (OPCS), of quantified food consumption. But there is less information about the extent of packaging use. It might be assumed that all food has been wrapped in a particular material. But this is an extreme assumption, although it is possible to achieve an estimate of intake using this approach. Clearly work is needed to define how much packaging and what types are used to wrap particular food commodities, so that estimates can be weighted for the amount of a wrapping material used according to its use in the market place.

Thus several methods of intake estimation can be used to assess the intake of a particular chemical contaminant for packaging but of course the legislator would wish to protect people with above-average intakes. For example it might be that brand loyalty causes a particular consumer only to buy food packaged in polystyrene or cans. Without taking account of this, an intake figure may be of limited use.

6.8 TOXICOLOGICAL ASSESSMENT

The toxicology of these substances is assessed, for legislative purposes, using data from a number of mutagenicity and other toxicological experiments which are beyond the scope of this chapter. However, it is worth noting that attempts have been made by computer to predict the toxicological effects of ingesting the chemicals. Such computer programs match the structure of the unknown compound with substances with well documented toxicity. One program based on quantitative structure-activity relationships (QSAR) was used to predict the likely toxicity of a range of compounds used in packaging. QSAR attempts to relate the biological properties of a compound with its physico-chemical properties. These physico-chemical properties include the electronic structure, the 3-D structure and the hydrophobic nature of the substance. Of these properties it has been found that there is good correlation between the electronic structure and the toxicity of, for example, phthalate esters. It is possible to predict the electronic structure using molecular orbital theory and to calculate electron densities of molecules using computer-based semi-empirical methods (MNDO, CNDO, MINDO/3 are acronyms given to the main types of molecular orbital calculations used in these predictions). Recent developments in computer software have shown how molecules might split into structural fragments, and have enabled possible metabolites to be identified. QSAR software could then be used to predict the toxicity of the fragments and metabolites. It would calculate their electronic structures and compare these to the electronic structures of known toxic substances. The results could provide an insight into any likely problems from the chemicals used in packaging. It could enable regulatory authorities to prioritize substances according to their likely toxicities. The difficulty with this method is that it requires a vast toxicity database to identify problem chemicals reliably.

In the European Community progress is being made to control the intake of chemicals from packaging. Directorate General III of the EC Commission is developing a positive list of all chemicals that would be allowed in plastic food packaging. The adoption of a chemical onto the positive list depends on a full assessment of the toxicological properties of the particular chemical and an understanding that the chemical is safe below a finite TDI value. Stringent guidelines on the testing and reporting procedures are laid down in order to ensure that the safety of the chemical can be adequately assessed. Thus stringent toxicological criteria have to be met before a substance can be allowed on a positive list. The result of this assessment procedure is usually the determination of a specific migration limit (SML) which is derived from the TDI by assuming that the average weight of a person is 60 kg and that consumption is of 1 kg of the food wrapped using a material containing the chemical. Thus the SML would be calculated as sixty times the TDI and expressed in mg/kg food. At present monomers used in producing plastics are mainly being evaluated but this work will be expanded to include other types of chemicals used in packaging materials, for example plastics additives.

6.9 CONCLUSIONS

Scientific work on chemical migration from food packaging has concentrated on plastics, with much less research on paper, board and glass. There has also been work on metals, but this has concentrated on a few areas, for example lead migration from soldered cans. One of the problems confronting workers in this area is the vast number of chemicals that needs to be considered. Nevertheless there has been considerable progress in the development and application of analytical methodology. This has helped legislators to build a framework for the control of monomer migration from plastic packaging but there remain many unknowns. These unknown factors might be addressed by techniques that are not used much in the study of most other chemical contaminants in food, for example the use of mathematical modelling, but this has yet to be established.

Advances in food wrapping have meant that our food is less likely to be contaminated with micro-organisms, but we must ensure that it is not chemically contaminated at unsafe levels from packaging.

FURTHER READING

Shepherd, N. J. *Food Chemistry*, **8**, 129–145. (1982)

Various authors. *Food additives and contaminants* 5i, ed. Ashby, R., and Vom Bruck, C. G. (1988)

Crosby, N. I., *Food packaging materials.* Applied Science, London. (1981)

Hotchkiss, J. H., *Food packaging interactions.* Amer. Chem. Soc., Washington DC. (1988)

Food contaminants: sources and surveillance. Creaser, C. and Purchase, R. (eds). RSC, Cambridge, UK. (1991)

Relevant EC Directives

The *Official Journals* (OJs) in which these are published are as follows:

76/893/EEC—OJ L 340, 9 December 1976.
78/142/EEC—OJ L 44, 15 February 1978.
80/590/EEC—OJ L 151, 19 June 1980.
80/766/EEC—OJ L 213, 16 August 1980.
81/432/EEC—OJ L 167, 24 June 1981.
82/711/EEC—OJ L 297, 23 October 1982.
83/229/EEC—OJ L 123, 11 May 1983.
84/500/EEC—OJ L 277, 20 October 1984.
85/572/EEC—OJ L 372, 31 December 1985.
86/388/EEC—OJ L 228, 14 August 1986.
89/109/EEC—OJ L 40, 11 February 1989 and corrigenda in OJ L 347, 28 November 1989, page 37.
90/128/EEC—OJ L 349, 13 December 1990 (corrected text).

7

Metals

N. Harrison, Ministry of Agriculture, Fisheries and Food, R238, Ergon House, c/o Nobel House, 17 Smith Square, London SW1P 3JR, UK.

7.1 INTRODUCTION

There is a continuing need to assess the contamination of the food supply by metals. It has long been recognized that the presence of elements known to be poisonous to humans such as lead and arsenic is undesirable in food and this has been promulgated in statutory legislation in the UK. In more recent years, other potentially toxic metals have come into focus. Cadmium and mercury have been the subject of much monitoring of the food chain and now other metals, in particular aluminium, are attracting attention. The UK's food surveillance programme on metals in food is overseen by the Working Party on Inorganic Contaminants in Food, formerly the Working Party on the Monitoring of Foodstuffs for Heavy Metals, under the direction of the Steering Group on Chemical Aspects of Food Surveillance (SGCAFS) which provides overall guidance to the Government's food surveillance in the UK (Chapter 9, section 9.7). A comprehensive programme of work has been established for a long time in this country such that the surveillance of the food supply should ensure that contamination of the food supply by metallic contaminants is kept to the minimum.

Surveillance of food for metals has provided a lot of data of use in estimating their dietary intakes. Among the methods available for the estimation of dietary intakes of heavy metals, three have been used extensively. These are: the total diet study; the diary study; and the duplicate diet study.

The UK Total Diet Study (TDS) relies on nationally representative information about the average food consumption by individual households researched in the UK National Food Survey (based on a survey of approximately 7000 households). Typical diets are constructed based on these data. Foodstuffs are purchased from retail outlets, then prepared and cooked in the normal manner. The individual foodstuffs are then usually combined into various groups of similar foods—for example cereals, green vegetables and fish—in the proportions eaten on average by consumers. Analysis of these food groups from various locations around the UK yields figures which, when combined with the consumption data, give the total intake of the contaminant by the average person for that year. The use of the TDS in this way provides information on the groups of food, if any, which are major contributors to dietary exposure. For example, most of the intake of arsenic and mercury is from the fish group, whereas lead is essentially a ubiquitous contaminant. Diary studies are used to determine in detail the consumption of a particular part of a diet. A population consuming above-average amounts of food that provides the main source of exposure to a contaminant can be identified using questionnaires. A record of the type and weight of food eaten, and the source, is then kept in a purpose-made diary by participants in the study. Representative samples of foods eaten are then analysed and the data combined. An extension of this approach is exemplified by the duplicate diet study, in which as exact a replicate as possible of all food consumed is collected for analysis. Further details of the generation of data on food intakes by these and other means are given in Chapter 10.

7.2 LEAD

Lead exists in small quantities in the earth's crust, but is well known as it is extracted easily from its ores. It is a grey, ductile, malleable metal which has been used extensively by man from earliest times, with references to it being found in Egyptian hieroglyphics of around 1500 BC. Amongst other uses, it was employed by the Romans for aqueducts and water mains. The possible hazard posed by the use of lead pipes was recognized in the first century BC by Vitruvius—it has even been suggested that the fall of the Roman Empire was hastened by the use of lead acetate to sweeten wine. Lead still has a number of important uses in the present day; from sheets for roofing to screens for X-rays and radioactive emissions.

The intake of lead in food by the general population in the UK is well within international tolerable limits. Results from the TDS indicate that during the period 1982 to 1987 the intakes over the whole population were between 0.02 and 0.07 mg per person/day (Table 7.1). This excludes the contribution from drinking water. (In the period 1975–1981 the mean dietary intake was estimated as 0.1 mg/day.) The average person in the UK is estimated to have a dietary lead intake of up to 0.42 mg/week; the Provisional

Table 7.1. Mean daily intake of lead (mg per person)

	1982	1983	1984	1985	1986	1987
Assuming lead levels in negative samples = LOD† (upper bound estimate)	0.07	0.07	0.06	0.07	0.06	0.06
Assuming lead levels in negative samples = zero (lower bound estimate	0.04	0.02	0.02	0.03	0.02	0.02

† LOD = limit of (analytical) detection (see Chapter 10, section 10.3.3.1).

Tolerable Weekly Intake (PTWI)† recommended by the Food and Agricultural Organization/World Health Organization (FAO/WHO) is 3 mg/week for a 60 kg person.

The distribution of lead between the various food groups for the 1988 UK TDS is given in Fig. 7.1. Lead concentrations in most foods were generally low, with all but a few samples containing lead at levels below the limits defined in the *Lead in Food Regulations, 1979*, as amended (there is a general limit of 1 mg/kg and limits for specific foods as defined in Schedules to the Regulations). Recent (1989) data on lead concentrations in individual foods are given in Table 7.2.

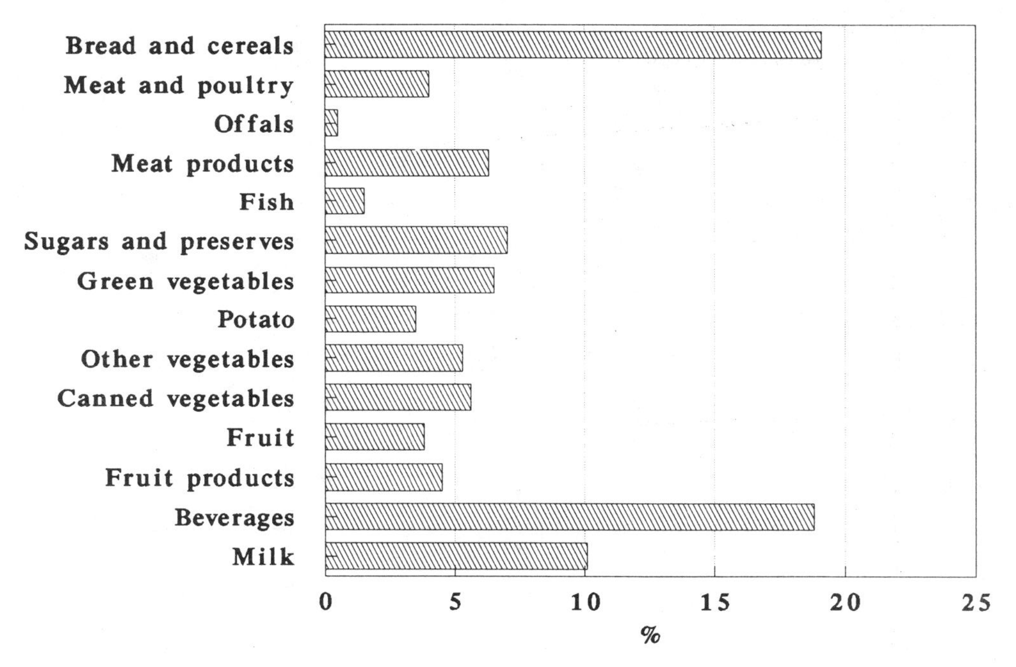

Fig. 7.1. Distribution of lead in UK total diet samples (1988).

Duplicate diet studies of people living in areas where exposure to lead from other sources is known to be low have provided some valuable information. The mean dietary lead intake of women was estimated to be 0.31 mg/week and the mean lead intake of children as 0.11 mg/week. These studies included the contribution from drinking water. In areas with elevated levels of lead in tap water, estimated lead intakes of both adults and children are found to be higher, and in a small percentage of cases above the PTWI.

For regular consumers of alcoholic beverages, wine and beer may make a significant contribution to dietary lead intake and this is reflected in the higher blood lead levels found for drinkers. Lead in beer is also absorbed more readily from the gastrointestinal tract than lead from the rest of the diet. A survey of lead in wines and beers from

† PTWI is analogous to ADI and TDI but is based on intake over a week.

Table 7.2. Lead in individual foodstuffs (mg/kg fresh weight)

Food	Number of samples	Mean (mg/kg)	Range (mg/kg)
Cereals			
Flour:			
Self-raising	8	< 0.05	< 0.05–0.05
Plain	8	< 0.05	—
Wheatmeal	3	< 0.05	—
Wholemeal	5	< 0.09	< 0.05–0.15
Bread:			
White	8	< 0.05	< 0.05–0.05
Wholemeal	8	< 0.05	—
Plain biscuits	5	< 0.05	—
Rice	5	< 0.06	< 0.05–0.10
Meat			
Beef (braising steak)	8	< 0.05	< 0.05–0.05
Lamb	8	< 0.05	< 0.05–0.10
Pork	8	< 0.05	< 0.05–0.05
Chicken	10	< 0.05	—
Other foods			
Concentrated soft drinks	14	< 0.01	< 0.01–0.01
Baby foods:			
cereal and dried	9	< 0.07	< 0.05–0.15
in jars	11	< 0.05	< 0.05–0.05
Chutney	5	< 0.08	< 0.05–0.20
Dried milk	10	< 0.06	< 0.05–0.10
Dried soup	5	< 0.05	< 0.05–0.05
Fish pastes	8	< 0.06	< 0.05–0.15
Stock cubes	5	< 0.12	< 0.05–0.35
Frozen foods			
Carrots	10	< 0.05	< 0.05–0.05
Spinach	11	< 0.05	< 0.05–0.05
Strawberries	4	< 0.03	< 0.02–0.06
Raspberries	4	< 0.02	< 0.02–0.02
Blackcurrants	4	0.05	0.04–0.06
Blackberries	4	0.04	0.02–0.08

Table 7.2. (continued)

Food	Number of samples	Mean (mg/kg)	Range (mg/kg)
Fresh fruit			
Pears	8	< 0.03	< 0.02–0.06
Apples	8	0.03	0.02–0.06
Tomatoes	2	< 0.02	—
Gooseberries	1	0.06	—
Loganberries	1	0.16	—
Blackberries	4	0.07	0.02—0.14

lead-capped bottles in 1982/1983 showed that about 90% of canned and bottled beers sampled contained < 10 μg/l and that nearly half the draught beers sampled contained > 10 μg/l, 4% contained > 100 μg/l. All wines sampled directly from the bottle contained < 250 μg/l.

The contribution from deposition of airborne lead on soil and crops to lead in diets is estimated to be between 13% and 31% for children. For individual plants a high percentage of lead may derive from aerial deposition (40–100%). Where crops are contaminated by lead from the air and soil, much of this may be removed by washing and other normal culinary practices.

Although dietary lead intakes in the UK are well within recommended intakes, it is the Government's policy to ensure that exposure to lead is reduced wherever practicable. Since 1982 action has been taken on a number of fronts:

— *Changes in canning technology* Changes by UK industry have resulted in the replacement of lead-soldered cans by using welded side-seams and this in turn has led to a fall in lead levels of UK-produced canned foods.
— *Contamination of draught beer by lead* This has been reduced by replacing dispensing equipment made from brass and other lead-containing alloys.
— *Lead levels in wine* Levels in wine poured from bottles sealed with tin-coated lead capsules have been found in some instances to be greater than 1 mg/kg. Action has been taken to phase out the use of these capsules. Retailers and consumers have been advised that all wine bottles should be wiped around the rim with a moist cloth after uncorking and before pouring so as to avoid undue contamination.

7.3 CADMIUM

Cadmium is naturally present in the environment: in soils, sediments and even in unpolluted seawater. It is closely related chemically to zinc and is found wherever zinc is also found. Thus most commercially-available zinc compounds will contain cadmium at low levels. Cadmium and its compounds have been used widely in industry with concomitant environmental pollution. Cadmium is emitted to air by mines, metal smelters and industries using cadmium compounds for alloys, batteries, pigments and in plastics.

The results of studies on animals show that cadmium is an extremely toxic metal, with renal damage being caused by long-term exposure. One sign of this damage is proteinuria (the appearance of increased levels of unfiltered proteins in the urine). Cadmium and its compounds can also give rise to carcinogenic effects in animals, although the evidence for similar effects in humans is not conclusive.

As cadmium is ubiquitous in the environment, all food is exposed to and contains cadmium. The main sources of cadmium contamination of the UK food supply are: phosphate fertilizers; atmospheric deposition; and sewage sludge.

There are currently no UK limits for cadmium in food. However, the recommendation of the UK Food Advisory Committee is that food containing levels of cadmium not acceptable in its country of origin should not be admitted to the UK.

In general, though the concentrations of cadmium in food in the UK are low, some foods of minor dietary importance such as shellfish or kidney often contain levels in excess of 0.5 mg/kg. Plant-based foods rarely contain more than 0.2 mg/kg on a fresh weight basis, although some root crops such as carrots and parsnip, and some leafy crops, such as spinach and lettuce, tend to contain more cadmium than other plant foods. This is also true of cereals, which indicates that plants tend to take up cadmium from the soil, unlike lead.

Most fish taken around the UK coast contain little cadmium, with the average being less than 0.2 mg/kg. The results of a routine surveillance exercise are given in Table 7.3. Shellfish contain higher concentrations than do most other foods, but, with the exception of lobster, whelks and crabs, shellfish from unpolluted waters rarely have an average cadmium concentration greater than 1 mg/kg. However, whelks and the body meat of lobsters and crabs may often contain higher concentrations.

Table 7.3. Cadmium in fish and shellfish (mg/kg fresh weight)

Fish	Number of samples	Mean cadmium concentration (mg/kg)
Bass	14	0.2
Cod	343	0.16
Herring	171	0.18
Monkfish	10	0.02
Whiting	360	0.17
Sea trout	68	0.03
Salmon	17	0.03
Shrimps	7	0.2
Winkles	1†	0.2
Whelks	4	1.8
Crabs (brown meat)	43	4.3
Crabs (white meat)	44	0.2

† Analysis made on 250 samples bulked together.

The level of cadmium in animal tissues, other than offal, is usually low, average concentrations being less than 0.05 mg/kg. Individual samples of kidneys may contain levels higher than 0.5 mg/kg. Animals grazing on land contaminated with cadmium will often have meat with normal levels of cadmium, while the level in offal is often significantly higher.

Table 7.4 presents the results from the TDS for cadmium between 1976 and 1988. The results show that for the average person in the UK, the dietary intake of cadmium has remained roughly the same over this period at approximately 18–20 μg/day. These are upper-bound estimates, that is in the estimation of dietary intake it is assumed that levels in food which are less than the limit of detection are equal to the limit of detection. These estimates therefore overestimate the actual intake to some extent.

Table 7.4. Mean daily intake of cadmium (μg/person)

1974	1975	1976	1977	1978	1979	1980	1981(a)	1981(b)
< 20	< 22	< 20	< 18	< 20	< 17	< 26	< 17	< 19

Note: These intake values were estimated on the basis that cadmium levels in samples in which no cadmium was detected, were equal to the limit of analytical detection.

Two sets of results are presented for 1981 because, in that year, some changes were made in the design of TDS food groups: (a) represents the estimated intake based on the analysis of the old total diet study food groups before the change took place, and (b) represents the estimated intake based on the new TDS food groups after the change had taken place. It can be seen from the table that the change in methodology did not result in any significant change in the estimated intake.

For adults, intakes of cadmium from food and other sources (e.g. air, water, smoking) can be compared with a provisional tolerable weekly intake (PTWI) of 400–500 μg, proposed by the Joint FAO/WHO Expert Committee on Food Additives (JECFA). The average weekly dietary intake of cadmium for adults in the UK is about 140 μg. However, the results of duplicate diet studies have shown that there are localized high intakes by consumers in certain areas or by consumers of certain foods. In the old mining village of Shipham in Somerset, UK, where the cadmium levels in some vegetable samples were more than 1 mg/kg, the cadmium intakes of the study sample were about double the average weekly dietary intake and some individuals exceeded the PTWI for cadmium. There was no evidence that any of the residents of the village had suffered adverse health effects related to cadmium. Further details of this study are given in Chapter 10 (section 10.4.5).

7.4 MERCURY

Mercury is ubiquitous throughout the environment. It occurs mainly in the form of the ore cinnabar (mercury (II) sulphide), but there are also at least thirty minerals in which the metal is found. Mercury is the only metallic element which is liquid at room temperature. Exposure to mercury vapour over periods of months or years can result in chronic poisoning. Intake of inorganic mercury compounds can result in bleeding from the gastrointestinal tract and kidney damage, followed by death from uraemia. Organomercury compounds have a wide variety of effects, and range from therapeutic agents to lethal chemicals. Of the organomercury compounds, the alkylmercurials are considered to be the most toxic. There have been a number of incidents involving alkylmercurials, of which methylmercury is the most intensively studied as it is thought to be the main agent of Minimata disease. In this incident, children of consumers of fish from Minimata Bay in Japan suffered from damage to the central nervous system, resulting in disturbances in vision, hearing, muscle function and mentality. It was later determined that there had been discharges of methylmercury in effluent from a chemical plant to the Bay.

There is no statutory control of mercury in food in the UK, but the Food Standards Committee have recommended that food which contained mercury at levels in excess of those acceptable in the country of origin should not be admitted into the UK.

The estimated daily intakes of total mercury (organic and inorganic) as determined from the UK TDS (Table 7.5) were between 0.010 and 0.014 mg/person in 1974, while more recent results (1982) show that the mean dietary intakes for the general population are between 0.002 and 0.003 mg/person per day. This may be compared with an ADI of 0.045 mg per day, for a 60 kg person, calculated from the JECFA PTWI.

Table 7.5. Mean daily intake of mercury (mg/person)

	1974	1975	1976	1977	1978	1979	1980	1982
Assuming mercury levels in negative samples = LOD† (upper bound estimate)	0.014	0.008	0.005	0.005	0.005	0.004	0.005	0.003
Assuming mercury levels in negative samples = zero (lower bound estimate)	0.010	0.003	0.004	0.003	0.004	0.003	0.004	0.002

† LOD = limit of (analytical) detection.

The highest concentrations of mercury are found in the fish group of the TDS, which contributes about 0.001 mg/kg to the overall intake. Higher than usual concentrations of mercury can sometimes be found in the cereals and the meat groups which indicates that contamination of these foods can occur.

Although the use of organomercury compounds as seed dressings is now no longer permitted in this country, it has been known for treated grain to be milled for human consumption in the past and there have been incidents in Iraq, Pakistan and Guatemala. A detailed consideration of the mercury concentrations found in the bread and cereals group of the TDS samples since 1976 reveals a very low level of contamination. A survey by the SGCAFS Working Party on the Monitoring of Foodstuffs for Heavy Metals in 1984 found that of 17 samples of the bread and cereals group analysed, none contained mercury in excess of 0.001 mg/kg. A hundred samples of grain analysed at the same time contained concentrations of less than 0.005 mg/kg in most cases, although four samples contained between 0.013 to 0.017 mg/kg total mercury. It is not known for sure why these levels were found, but similar concentrations have been known to accumulate in grain due to atmospheric contamination.

Mercury has been shown to accumulate in eggs when organomercury compounds are present in cereal used as feed. Egg white normally contains higher levels of total and organic mercury than the yolk owing to an association of mercury with the protein ovalbumin. In 1984, 75 samples of retail chicken eggs were analysed for total mercury and inorganic mercury. The mean mercury concentration was about 0.004 mg/kg (range < 0.0005 to 0.029 mg/kg). An extreme consumption of eggs would be about 500 g/week (about one-and-a-half eggs a day) and taking the mean figure results in a mercury intake of 0.01 mg/week.

Mercury is normally present in fish as it is ubiquitous in the environment. Fish and shellfish caught off the coasts of England, Wales and Scotland have been regularly monitored for mercury since 1973; the most recent results are for 1990. These data are used to meet UK obligations in the European Community and to the Oslo and Paris Conventions. An example of these data are those collected for Liverpool Bay, into which the Mersey flows. The Mersey is subject to discharges (from chlor-alkali plants) some of which can contain mercury. Data are submitted annually to the EC to indicate compliance with the Environmental Quality Standard of 0.3 mg/kg mercury wet weight in fish flesh. In recent years there has been a significant reduction in the amount of mercury discharged into the marine environment: the data for 1977–1984 show a 52% reduction in mean mercury concentrations in all species of fish in the Liverpool Bay area. The mean mercury concentrations decreased from 0.23 to 0.20 mg/kg between 1985 and 1987. Thus it is apparent that, overall, levels of mercury in fish in Liverpool Bay are decreasing.

As it is known that fish is a major source of mercury in the diet, extensive surveillance work has been carried out. Predatory species such as tuna and swordfish tend to accumulate relatively high levels. During 1973–74, a survey conducted by MAFF determined that mean mercury concentrations in imported fish were in general less than 0.1 mg/kg, although halibut and tuna were exceptions with mean concentrations of 0.16 mg/kg and 0.12 mg/kg respectively. Most of the species analysed had concentrations of less than 0.05 mg/kg. A further survey of frozen, tinned or dried imported fish in 1983 showed that with the exception of tuna and one sample of sardines, levels were all below 0.05 mg/kg, confirming the earlier findings. The mean level of tuna (tinned) was

0.12 mg/kg (range: 0.04 mg/kg to 0.22 mg/kg). Although this level was higher than those for the other species of fish, it is similar to the concentration found in the earlier survey indicating no large change in mercury concentrations.

7.5 ARSENIC

Although arsenic is a metalloid, it is generally included in work on metals in food. It is rarely found as the free element in the natural environment, but more commonly as a component of sulphur-containing ores in which it occurs as metal arsenides. Arsenic is present in rocks, soils, water and living organisms in concentrations of up to parts per million (mg/kg).

Interest in the contamination of food with arsenic arose from a serious outbreak of arsenical poisoning in Northern England in 1900 due to beer made with glucose. It was established that the outbreak arose chiefly as a result of the use of arsenical pyrites to manufacture sulphuric acid used in production of the glucose. A Royal Commission set up in 1903 to consider the case concluded that maximum levels of arsenic in food and liquids should be established. The general limit for food is 1 mg/kg, and 0.1 mg/kg for ready-to-drink beverages. Three categories of food are specifically excluded:

— naturally present arsenic in fish and edible seaweed or products containing fish or edible seaweed;
— hops or hop concentrates intended for commercial brewing; and
— any food with an arsenical level limited by other regulations.

The weekly intake of arsenic from food can be calculated from an analysis of the various food groups of the TDS. The most recent data are for 1974–82. The major source of arsenic in the diet is fish which, in 1982, accounted for nearly 75% of the total intake from food. The weighted average of the intakes from all the UK Total Diet Studies is 89 μg/day or 0.62 mg/week.

The mean concentration of arsenic in samples of fresh marine fish caught in UK coastal waters in 1967–1977 ranged between 0.7 mg/kg and 13 mg/kg. It was found that fish which live on or close to the sea bed, such as plaice, dabs, flounders and skate, tend to have higher levels of arsenic than other fish, but a level of 10 mg/kg is rarely exceeded. Arsenic levels in shellfish show more variation. High levels are frequently found in crab, in which the white meat generally contains more arsenic than the brown meat. Lobsters contain lower levels of arsenic than crabs. An important consideration when assessing the significance of these elevated levels is the finding that arsenic is almost entirely present in fish as the organic chemical arsenobetaine, which has been found not to be metabolized in man.

The total, that is inorganic plus organic, arsenic content of dietary supplements has been determined. Samples of kelp powder and tablets were analysed in 1987 for both total arsenic and inorganic arsenic (unstable organic species were also included). Levels from 7 to 45 mg/kg of total arsenic were found, with an outlying value of 120 mg/kg. The inorganic arsenic content was between 0.01 to 0.45 mg/kg, with the same outlying sample containing 50 mg/kg.

Other than fish, meat products are the main food where arsenic has been consistently detected, with levels of up to 9 mg/kg in pig liver. These levels were detected at a time when organoarsenical feed additives were permitted as growth promoters.

7.6 ALUMINIUM

Aluminium is the third most abundant element in the earth's crust and is used widely in the manufacture of construction materials, wiring, packaging materials and cookware. The metal and its compounds are used in the paper, glass and textile industries as well as in food additives. Despite the abundance of the metal, its chemical nature effectively excludes it from normal metabolic processes. This is due largely to the low solubility of aluminium silicates, phosphates and oxides which result in the aluminium being chemically unavailable. However, it can cause toxic effects when there are raised concentrations of aluminium in water used for renal dialysis. These effects are not seen when aluminium is at the concentrations usually present in drinking water.

In order to assess the contribution of metals leaching into canned foods, a special Total Diet Study was set up by MAFF in 1984 with diets of the following compositions: as normal; as normal but containing no canned food; and as normal but containing canned food wherever possible.

The results of this study (Table 7.6) show that canned food does not contribute above-average amounts of aluminium to the diet.

Table 7.6. Aluminium intakes from whole diets containing different amounts of canned foods

Diet	Proportion of canned food (% w/w)	Intake (mg/day)
Normal	4.6	6
No canned food	0	7
High content of canned food	39	5

Note: The normal Total Diet contains 6.2% w/w of canned food; the prepared 'normal' diet contained 4.6% w/w canned food.

Aluminium has been measured in a range of crops. A large variation in levels is apparent for the various species. The highest concentrations were found in lettuce (6.2–810 mg/kg), parsnips (6.0–82 mg/kg) and Brussels tops (7.7–116 mg/kg). The intake for the 1988 TDS was 4 mg/person per day, with plant foods being the main dietary source. This figure may be compared with the PTWI for aluminium of 7 mg/kg bodyweight which is equivalent to 60 mg/day for a 60 kg adult. There is currently much activity to examine the factors which influence uptake of aluminium from the diet.

7.7 OTHER METALS

Most of the surveys on metals in food have concentrated on those metals which are known to be toxic, or where there are possible concerns about their levels in food. In the course of collecting the data, information on other metals is often collected in addition. Other metals which have been included in the UK Government's surveillance are copper, zinc, antimony, chromium, cobalt, indium, nickel, thallium and tin.

7.7.1 Copper

Copper is an essential element for all plants and animals. It is widely distributed and always present in food. Animal livers, which are the major contributor to dietary exposure to copper, various shellfish and some dry materials contain, on average, more than 20 mg/kg. This was the limit recommended for food by the Food Standards Committee in 1956 and is also a statutory limit for tomato ketchup. Milk contains little copper, usually less than 0.1 mg/kg, and fresh fish and alcoholic drinks contain less than 1 mg/kg of copper.

The mean daily intake of copper from UK Total Diet Study (TDS) samples fell from 2.3 mg/person in 1972 to < 1.6 mg/person in 1978.

7.7.2 Zinc

As with copper, zinc is an essential element for all plants and animals. It is necessary for the correct function of various enzyme systems. However, excessive intakes of zinc can have long-term effects. In food, the major contributors to the diet are meat and its products, from which zinc is readily absorbed. Liver, with concentrations of around 62 mg/kg, contains the highest levels of any meat products, with other tissues having values of a half to a third of this figure. The second greatest source of zinc is cereals. Concentrations of zinc in whole cereal products are similar to those in meat.

Over the period 1972 to 1978, the estimated mean zinc intake in the UK varied from 8.8 to 11.5 mg/person per day.

7.7.3 Antimony

Compounds of antimony are used as fire retardants in plastics and paper, and for veterinary purposes. The metal is found in specialized alloys such as white metal bearings and pewter—which is an alloy of tin, antimony (up to 7.5%) and copper. Concentrations in food are low, generally in the range < 0.01 to 0.08 mg/kg, but have been found to be higher in samples of aspic jelly and cream of tartar. The estimated daily intakes from the UK TDS were estimated to be 2 to 29 μg/person per day in 1985.

7.7.4 Chromium

Chromium is used in the manufacture of stainless steel and other specialist steels, and non-ferrous alloys. Chromate salts are used as tanning agents, pigments, catalysts, corrosion inhibitors and in electroplating solutions. Although stainless steel is inert, the dissolution of chromium from this steel is likely to be the major source of chromium in food.

Chromium is an essential element for man. The minimal requirement for man is estimated to be about 1 μg/day. As the absorption of inorganic chromium (Cr III) is about 0.5% of a given dose, and the absorption of organically-bound chromium is even higher, a dietary intake of 200 μg Cr/day will provide the estimated requirement. A chromium intake of 50 to 200 μg/day has been recommended for adults by the US National Academy of Sciences.

From the TDS, the weighted mean intake for 1976 to 1978 was 136 μg/person per day (upper bound) and 111 μg/person per day (lower bound) (Table 7.7).

Table 7.7. Mean daily intake of chromium (μg/person)

	1976	1977	1978
Assuming chromium levels in negative samples = LOD† (upper bound estimate)	130	172	99
Assuming chromium levels in negative samples = zero (lower bound estimate)	94	155	77

† LOD = Limit of (analytical) detection.

7.7.5 Cobalt

Cobalt is used in the manufacture of alloys and in nuclear technology. Cobalt compounds are included in trace element supplement preparations for ruminants. The cobalt concentrations in vegetables and other foods has been found to be between < 0.01 and 0.83 mg/kg, while levels in milk are between 0.0002 and 0.06 mg/kg.

The mean daily intakes of cobalt estimated from TDS samples over the period 1977 to 1978 are given in Table 7.8.

Table 7.8. Mean daily intake of cobalt (μg/person)

	1977	1978
Assuming cobalt levels in negative samples = LOD† (upper bound estimate)	27	28
Assuming cobalt levels in negative samples = zero (lower bound estimate)	11	14

† LOD = Limit of (analytical) detection.

7.7.6 Indium

Indium occurs as an impurity in tin ores and is added to some alloys to impart corrosion resistance and surface hardness. It is also used in the production of photovoltaic cells. It is poorly absorbed from the gastrointestinal tract, but some absorption does occur. In crops it has been found at levels of between < 0.01 and 0.04 mg/kg. A 1979 TDS determined the estimated daily intakes of indium from the diet to be between 5 μg/person and 27 μg/person.

7.7.7 Nickel

Salts of this metal are widely used in industry for nickel plating and as pigments. There are thought to be at least 3000 nickel alloys which have usage in storage batteries, coins, cooking utensils and other products. Fats and oils are hydrogenated using nickel as a catalyst.

The highest concentrations of nickel in individual foods occur in tea, soya protein and herbs. Other foods with elevated levels are pulses, cocoa products and some nuts. Nickel has been found in crops at levels of < 0.01 to 1.5 mg/kg, with the highest concentrations in broad beans and Brussels sprouts. The mean daily intakes of nickel estimated from TDS samples over the period 1974 to 1981 are given in Table 7.9.

Table 7.9. Mean daily intake of nickel (mg/person)

	1974	1976	1977	1978	1980	1981
Assuming nickel levels in negative samples = LOD† (upper bound estimate)	0.21	0.33	0.27	0.28	0.27	0.23
Assuming nickel levels in negative samples = zero (lower bound estimate)	0.16	0.32	0.26	0.27	0.26	0.22

† LOD = Limit of (analytical) detection.

7.7.8 Thallium

Thallium salts are known to be toxic to the gastrointestinal tract and the central nervous system. This rare metal and its compounds find uses in various industrial applications, such as in the manufacture of pigments, dyes, artificial gems and window glass. The concentrations of thallium in food in general in the UK are less than 0.05 mg/kg, and lower in beverages. The estimated daily intakes of thallium from the diet in 1979 were between 5 and 61 μg/person per day.

7.7.9 Tin

Tin is widely distributed in nature at low concentrations in marine and land plants, and animals. It has extensive uses in industry, including in pewter, and its compounds are used as heat stabilizers (organotin) in plastics, in glazes for porcelain (salts), and coatings (oxide) to reduce abrasion of glass containers. The main source of tin in food is tin-plated steel used in the manufacture of cans for foods and beverages.

Concentrations of tin in most foods are less than 1 mg/kg, although higher concentrations have been found in cereals. The mean daily intakes of tin estimated from TDS samples over the period 1976 to 1981 are given in Table 7.10.

Table 7.10. Mean daily intake of tin (mg/person)

	1976	1977	1978	1979	1981
Assuming tin levels in negative samples = LOD[†] (upper bound estimate)	4.4	4.2	3.6	3.2	2.4
Assuming tin levels in negative samples = zero (lower bound estimate)	4.2	4.1	3.5	3.1	2.3

† LOD = Limit of (analytical) detection.

The fall in tin intake shown in Table 7.10 is likely to be owing to a change to lacquered food cans. The major source of dietary tin is canned food, the metal entering the food by dissolution of the tin plating, which lacquering can reduce. The results of a recent survey of canned foods monitored over the period 1983 to 1987 showed that the tin concentrations in the canned food were generally below 200 mg/kg. Less than 2% of the cans sampled contained food with levels in excess of 200 mg/kg. Tomato-based products tend to have high levels of tin as nitrate in the food accelerates corrosion of the tin.

FURTHER READING

Alloway, B. J. (ed.) *Heavy metals in soil*. Blackie, Glasgow and London. (1990)

Berman, E. *Toxic metals and their analysis*. Heyden and Son, London. (1980)

Lead

Food Surveillance paper No. 27. HMSO, London. (1989)

The Lead in Food Regulations. S.I. No. 1254 (as amended). (1979)

Lenihan, J. and Fletcher, W. W. (eds) *The chemical environment*. Blackie, Glasgow and London. (1977)

Cadmium

Mislis, H. and Ravera, O. *Cadmium in the environment*. Birkhäuser Verlag, Basel, Boston and Stuttgart. (1986)

Food Surveillance Paper No. 12. HMSO, London. (1983)

Mercury

Food Surveillance Paper No. 17. HMSO, London. (1983)

Aquatic environment monitoring report, No. 30. MAFF, London. (1992)

Arsenic

Food Surveillance Paper No. 8. HMSO, London. (1982)

The Arsenic in Food Regulations. S.I. No. 831 (as amended). (1959)

Norman, J. A., Pickford, C. J., Sanders, T. W. and Waller, W. *Food additives and contaminants*, **5**, 103–109. (1988)

Aluminium

Sigel, H. and Sigel, A. (eds). *Aluminium and its role in biology.* Vol. **24** Metal ions in biological systems. Marcel Dekker, New York and Basel. (1988)

Food Surveillance Paper No. 15. HMSO, London. (1985)

Miscellaneous additives in food regulations S.I. No. 1834. (1980)

Massey, R. C. and Taylor, D. (eds) *Aluminium in food and the environment.* Royal Society of Chemistry, London. (1989)

Copper and zinc

Food Surveillance Paper No. 5. HMSO, London. (1981)

The Food Standards (Tomato Ketchup) Amendment Regulations. S.I. No. 1167. (1956)

Antimony, chromium, cobalt, indium, nickel and thallium

Food Surveillance Paper No. 15. HMSO, London. (1985)

Tin

Food Surveillance Paper No. 15. HMSO, London. (1985)

The Tin in Food Regulations. S.I. No. 496. (1992)

Meah, M. N., Smart, G. A., Harrison, A. J. and Sherlock, J. C. *Food Additives and Contaminants*, **8**, 485–496. (1991)

8

Pesticides residues

D. H. Watson, Ministry of Agriculture, Fisheries and Food, R242, Ergon House, c/o Nobel House, 17 Smith Square, London SW1P 3JR, UK.

8.1 INTRODUCTION

Unlike many of the other chemical contaminants that are reviewed in this book there seems to be considerable public interest in the risks that pesticides residues might pose to the consumer. Therefore there is special reason to be clear about which pesticides are likely to give residues in food, how this happens, what controls are applied and what scientific work is carried out to ensure that the controls are effective. These topics are reviewed in this chapter, using information from UK scientific work to illustrate the main points. Work in this country is extensive, both pre-approval research on new pesticides and the surveillance of pesticides residues in food supply. There is also an important international dimension to work on pesticides residues. For example both the European Community and the United Nations' Codex Alimentarius Commission have set Maximum Residue Limits (MRLs) for pesticides residues in food, drawing on the results of research and food surveillance in many countries in doing so. There is also continuous international review of the toxicological Acceptable Daily Intakes (ADIs) for pesticides residues. These international aspects are also considered in this chapter.

Scientific work on pesticides residues in the diet started in the 1960s, before much of the science described elsewhere in this book had begun to develop. (The one exception is work on metals in food which has also been carried out for many years—Chapter 7.) Thus many of the approaches developed in work on pesticides residues have been used in work on other chemical contaminants, especially organic chemical contaminants. In particular there has been a transfer of technology used in analysing food for pesticides at their usually very low levels in food, to help analysis of other chemical contaminants in the diet. Thus scientific work on pesticides residues is of general importance to work on chemical contaminants in food. In particular there are many parallels between work on pesticides residues and studies on veterinary drug residues. Since the latter group of substances is reviewed in detail in Chapter 2, only the main points about pesticides residues are reviewed here.

8.2 DIFFERENT PESTICIDES

Pesticides can be classified in a number of ways—by their chemical structures, by their uses, by their chemical properties, and so on. Because work on pesticides residues is largely chemical in its nature, pesticides residues are usually described by their chemical names and classified by the groups of chemicals to which they belong, for example 'the organobromine compound, methyl bromide'. Several groups of pesticides are commonly included in residues work, notably insecticides. The main groups of insecticides are as follows:

— *Organochlorine pesticides* (Table 8.1) These are mainly the classically persistent insecticides, of which DDT is perhaps the best known example. These pesticides are now much less used worldwide than they were in the 1960s and 70s, because less persistent alternatives have been developed (see next page). Nevertheless they are detected in food samples and human body fat, although their levels are decreasing gradually in the environment in many developed countries. Section 3.3.2 of Chapter 3 discusses those organochlorine pesticides which probably now enter the food chain mainly from the environment.

Table 8.1. Some organochlorine pesticides (uses are not necessarily those approved, if any, in the UK)

Name	Structure	Uses
Dieldrin	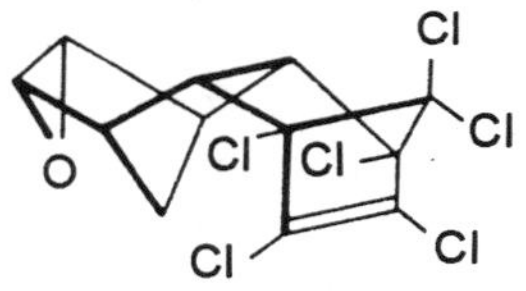	As an insecticide, for example against termites and locusts.
pp'-DDT	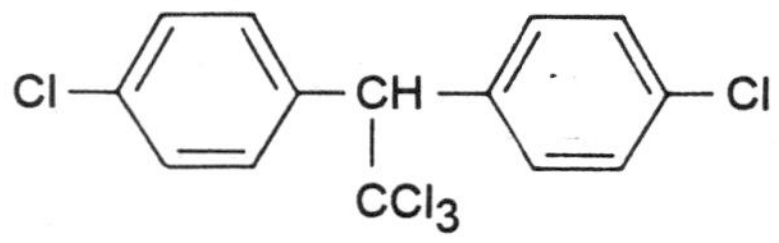	Insecticide (much less used than in the past).
gamma-HCH	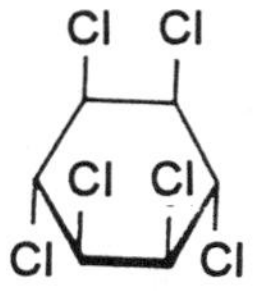	Insecticide (little or no use in many countries).
Pentachlorophenol	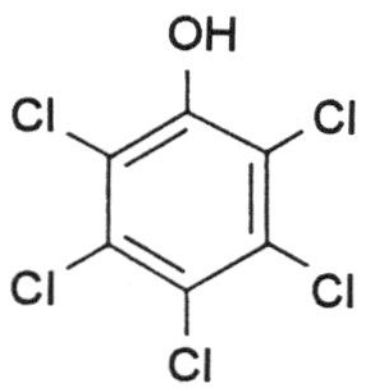	Wood preservative, pre-harvest defoliant and general herbicide.

— *Organophosphorus pesticides* (Table 8.2) These are mainly used as insecticides. They include chlorinated compounds, such as chlorpyrifos, which can be included in either this or the organochlorine group of pesticides. (Such chlorinated organophosphorus compounds are usually classed as organophosphorus compounds however.) Organophosphorus pesticides have been increasingly used instead of the organochlorines in many ways over the past decade or so, and are less environmentally persistent.

Table 8.2. Some organophosphorous pesticides (uses are not necessarily those approved, if any, in the UK)

Name	Structure	Uses
Chlorpyrifos	Pyridine ring with N; substituents Cl, Cl, Cl and $OP(OCH_2CH_3)_2$ with $=S$	Insecticide (e.g. control of mosquitoes, crop pests in soil and on foliage).
Diazinon	Pyrimidine ring (N, N) with CH_3, $CH(CH_3)_2$ and $OP(OCH_2CH_3)_2$ with $=S$	Insecticide (e.g. against external parasites on animals, and flies in glasshouses).
Malathion	$(CH_3O)_2P(=S)SCH(CO.OCH_2CH_3)CH_2CO.OCH_2CH_3$	Insecticide and acaricide (i.e. to kill mites).
Pirimiphos-methyl	Pyrimidine ring (N, N) with CH_3, $N(CH_2CH_3)_2$ and $OP(OCH_3)_2$ with $=S$	Insecticide and acaricide.

— *Organobromine pesticides* This is a small group, the main examples being methyl bromide (CH_3Br) and ethylene dibromide ($BrCH_2CH_2Br$). These two substances are used in some countries to fumigate soil and stored crops.

— *Pyrethroid pesticides* (Table 8.3) This group contains some compounds which are also organochlorine compounds (e.g. permethrin), but like chlorine-containing organophosphorus pesticides the chlorinated pyrethroids are not usually considered with classical organochlorine pesticides such as DDT.

The use of chemical structures to classify pesticides does not work quite so well for other groups of pesticides such as fungicides. For example four fungicides that are quite commonly included in residues work are listed in Table 8.4. Two of these, quintozene and tecnazene, are very closely related. But the other two compounds, captafol and bitertanol have quite different structures. The structures of herbicides are also diverse.

The existence of several chemically distinct groups of insecticides has helped studies on their residues in food, because it has allowed the production of methods of analysis for groups of these substances. Thus residues of organochlorine pesticides are generally measured by one common procedure. This has also allowed the detection of non-pesticidal organochlorine compounds, such as polychlorinated biphenyls, by the same method of analysis (see Chapter 3, section 3.3). This type of approach has also allowed the analysis of many samples of food and other materials for each of the above groups of insecticides. As a result, and because the analytical work for organochlorine pesticides started in the 1960s, there is a great amount of data on residues of many insecticides in food and food raw materials. Studies on the other pesticides in food samples are generally more recent and less extensive, although there has been quite intensive study of some fungicides in food, for example on the substances listed in Table 8.4.

Although there are several different categories of pesticides and many compounds in some of the categories, relatively few pesticides are likely to be found in food. There are several reasons why pesticides may not form residues in food:

— Not all pesticides are sufficiently persistent. The pesticides shown in Tables 8.1 to 8.4 persist to varying degrees in the food chain, but they are all sufficiently persistent to lead to residues in food. They might be thought of as being 'inherently' persistent. This is not the case for many pesticides.
— Certain pesticides, such as some weed killers, are sufficiently toxic to plants that they will kill any crops they may come into contact with, rather than leave residues in a crop that can be harvested.
— Not all pesticides are used in food production or storage. Although some may claim that this could still lead to residues as a result of adventitious contamination, this implies very weak control over pesticide usage which is not the case in many countries. Nevertheless it is important to keep a watch over such possible contamination of food, and this forms part of the food chemical surveillance work (Chapter 9) in many developed countries.

It is essential that the potential for residues formation is studied intensively and the resulting experimental data reviewed for any new pesticides. This is indeed the case for pesticides used in some countries, including the UK where the potential for residues in the food chain and the environment is carefully reviewed on the basis of controlled experimental trials in deciding whether a pesticide can be approved for use. There are similarly tough controls in the licensing of pesticides which are used medicinally on

Table 8.3. Some pyrethroid pesticides (uses are not necessarily those approved, if any, in the UK)

Name	Structure	Uses
Permethrin	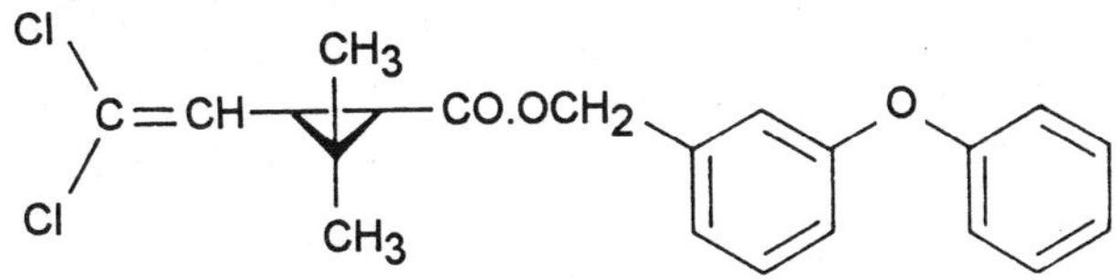 Cl, Cl, C=CH, CH_3, CH_3, $CO.OCH_2$, O	Insecticide (e.g. against external parasites, on animals, and leaf and fruit eating pests).
Cypermethrin	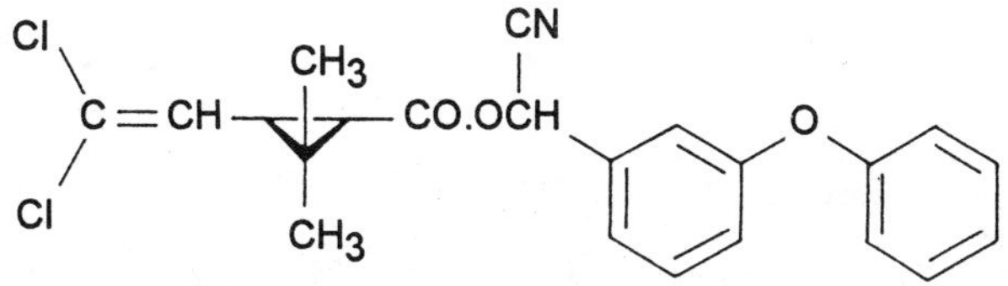 Cl, Cl, C=CH, CH_3, CH_3, CO.OCH, CN, O	Insecticide (particularly to kill *Lepidoptera*).
Fenpropathrin	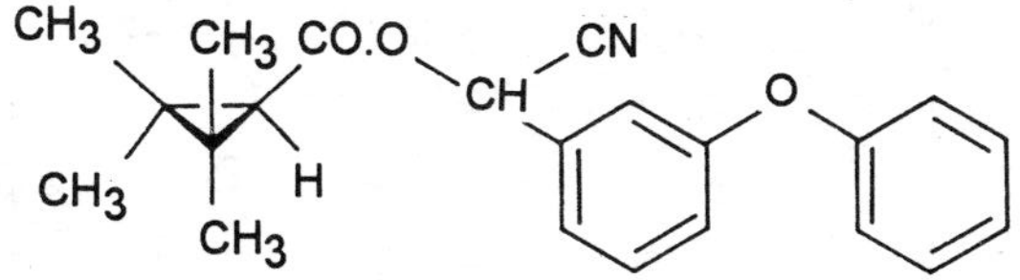 CH_3, CH_3, CH_3, CH_3, H, CO.O, CH, CN, O	Acaricide and insecticide.

Table 8.4. Some fungicides

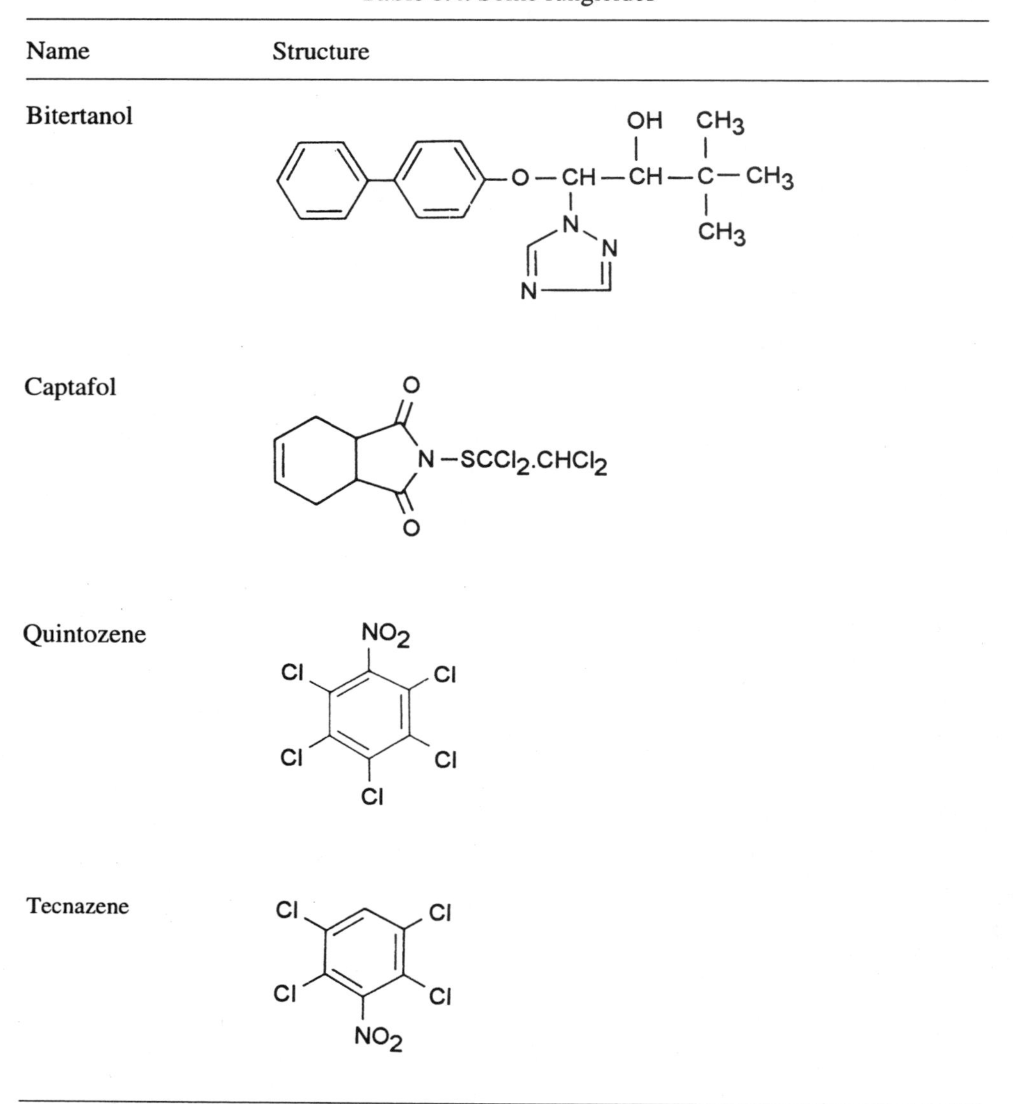

Name	Structure
Bitertanol	
Captafol	
Quintozene	
Tecnazene	

animals in this country (Chapter 2). To double check on the effectiveness of these controls, and on residues in imported foodstuffs, there is extensive food surveillance (Chapter 9) for pesticides (and veterinary drug) residues in the UK and several other countries, for example other EC Member States, the USA, Australasia and Canada. Food chemical surveillance provides a picture of the effectiveness or otherwise of controls, as well as of our exposure to residues of pesticides in the diet.

8.3 RESIDUES IN FOOD

8.3.1 Origins

Pesticides can enter the food chain at virtually any stage between crop production or animal husbandry on the farm, and consumption of food in the home. These chemicals are unique in the variety of possible ways in which they can contaminate food. Not even environmental chemicals (Chapter 3) seem to follow so many routes. The types of routes by which pesticides can reach the food chain include the following:

— *Treatment of crops* This includes the use of pesticides to kill insects, fungi and other pests, and preventative action to avoid infestation. Usage can be on growing crops or on harvested products, such as stored grain.
— *Veterinary usage* Insecticides are mainly used to treat or prevent illness caused by a wide variety of insects.
— *In food manufacture and retailing* Here pesticides are used mainly to prevent rodent and insect infestations. Since this type of treatment is of the environment in which food is produced or sold rather than of the food itself, it is likely to provide less input of pesticide residues into food—provided all other factors are equal (e.g. if the pesticides used in the food factory are of similar persistence to those used on crops or farm animals).
— *In the general environment* From time to time food surveillance detects residues of a pesticide that is not used in food production or storage, and investigation finds another source such as heavy industry or transport. However, there are relatively few pesticides that find such specialist use and these substances are uncommon contaminants of the food chain because they are not very persistent in the environment. An exception is some of the organochlorine pesticides which are highly persistent. Although they are generally no longer used in food production, residues are still found in food because of their persistence in the environment.
— *In the home* Just as in food manufacture the use of pesticides in houses can lead to indirect contamination of food. Clearly care is needed wherever food is being processed—in the factory and in the home—to avoid any possible contamination with pesticides.

Since there are so many different ways in which the usage of pesticides can lead to residues in food, it is often impossible to judge with any certainty where a particular residue has come from. Clearly where a pesticide has one main specific use it is possible to trace back and check that this was the cause of a residue. For example some organophosphorus pesticides are used mainly to protect stored grain from insects, and where they are detected in cereal products the source is fairly obvious. However, many other pesticides are used in several different ways, and in this circumstance considerable investigative work may be needed to identify the main source or sources of their residues in a given foodstuff.

8.3.2 Control

It is important to be able to identify the sources of pesticides residues because without this information it is very difficult to control the safety of the food supply. The most effective control methods are likely to be those which apply at the source of contamination, if only because control at a later stage almost inevitably means greater loss of food from the general supply. (Fig. 8.1 shows some possible control points.) Because pesticides can be used at all stages in the food supply chain, it is essential to have control points at each stage. In many countries, including the UK, this is the case. Historically controls in most countries tend to have been placed first on the usage of pesticides, but it is difficult to control residues solely in this way. Hence the other types of controls noted in Fig. 8.1 have also been introduced. For example in the UK there are legal powers to control maximum residue levels of pesticides in food. Clause 16(2) (k) of Part III of the *Food and Environment Protection Act 1985* says that 'Ministers may ... by regulations ... specify how much pesticide or pesticide residue may be left in any crop, food or feeding stuff'. Similar powers are available in most other countries in the developed world. Thus in many countries control over pesticides can be applied at several stages, from their usage to their presence in food on sale.

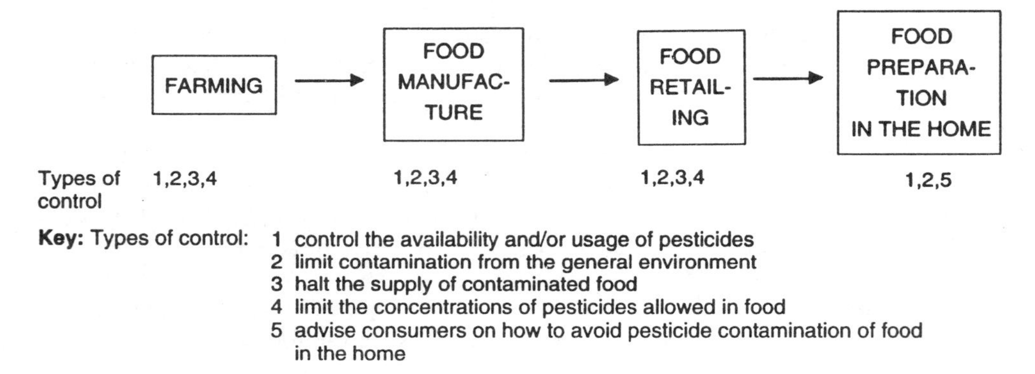

Fig. 8.1. Stages in food production at which pesticides residues might be controlled.

The control of pesticides residues in food is the responsibility of all involved in food production. This includes farmers, the food industry, and government. The emphasis in the UK and many other countries is on the prevention of pesticide contamination as much as on controlling excessive levels of residues. Where pesticides must be used then their levels have to be controlled, but where reasonable changes in industrial or consumer practice can avoid pesticides residues in food this opportunity must be taken. This has led to research to develop less non-persistent biological pesticides, for example natural predators of pests. However, this type of approach can be expensive compared with the use of persistent pesticides, and so especially where undernourished populations need to be fed as cheaply as possible the use of older persistent pesticides may occur. This provides a potential problem not only for the endemic population but also for countries receiving exports from such parts of the world. Thus in many developed countries there are also controls on the maximum levels of pesticide residues in imported food.

These major influences on the control of pesticides residues have led to the definition of internationally agreed maximum residue limits (MRLs) for many pesticides in food. Such MRLs are defined by the United Nations' FAO/WHO Codex Alimentarius

Commission. The same body also recommends acceptable daily intakes (ADIs) for pesticide residues (Chapter 10, section 10.6). However, unlike for veterinary drug residues where MRLs and ADIs are both linked to the toxicology of the chemical, Codex MRLs for pesticide residues are developed on the basis of the maximum residue that should be left after good agricultural practice—this may give a lower MRL than one based on health grounds. Other international bodies have also developed MRLs for pesticides residues, notably the European Community. They have not always used the same approach as the Codex Committee and so there is now more than one MRL for a given pesticide in some cases. In practice this has not caused major problems in international trade, however.

8.3.3 Detection

Analysis of food for pesticides has been carried out for many years, with the result that there are many different methods available for the extensively studied compounds, notably the organochlorine insecticides (section 8.2). As noted before the use of methods to detect several pesticides via the same analytical method has led to extensive data on pesticides residues in food and to the analysis of non-pesticides, such as polychlorinated biphenyls, by the same method that is used to detect organochlorine pesticides. There are now 'multi-residue' methods for several groups of pesticides. Such methods must be capable of detecting and quantifying the residues at very low concentrations in food (generally below one part per million), because MRLs are usually set at very low levels.

The main steps involved in analysing food for pesticides residues are the same as those in analytical work on any chemical contaminant in food (see Chapter 9, section 9.2). However, the use of multi-residue methods requires the application of highly effective methods of separation, detection and confirmation, as the groups of pesticides are so similar in their chemical properties and are usually present in very small amounts. This has led to the use of a number of different techniques, including gas chromatography coupled with mass spectrometry, using two different chromatography columns with specific detectors to identify groups of pesticides, and employing mini-computers to compare the characteristics of a given chromatography peak with information about different pesticides in the computer's database. But these types of approach can be very expensive. Another advance has been to use information about the different retention times of pesticides in various chromatographic systems. Although this is no replacement for chemical characterization of pesticides residues, it has greatly eased routine identification and hence allowed greater application of multi-residue methods. Nevertheless there is much debate amongst pesticides residues analysts, for example about the optimum conditions for using a particular type of column in multi-residue methods for analysing pesticides residues. As in other areas of chemical contaminant analysis there is also increasing effort to ensure effective analytical quality control—collaborative trials involving different laboratories are becoming increasingly common. Wherever residues data are reported it is imperative to ask what analytical quality assurance applies. Indeed it is also essential to check on the effectiveness of the other activities involved in residues work, for example how sampling was carried out, when pesticides residues data are considered.

8.3.4 Surveillance

Of all the organic chemical contaminants in food, pesticides residues have almost certainly been surveyed most extensively. Surveillance is mainly by the analysis of food samples. There are also surveys of pesticide usage which help in identifying potential problems with residues in food. Surveys which involve the analysis of samples for pesticides are carried out in the following main ways.

— *Overall dietary surveillance* in which a composite diet is prepared and analysed. This is a very 'broad brush' technique but it has the advantage of giving a picture of overall exposure via the diet to a given pesticide or group of pesticides. This can be very useful where a rough measure of a pesticide's intake is needed, for example to allow comparison with an ADI (sections 8.3.2 and 9.5). The UK total diet study is one example of this type of approach (see Table 9.5).

— *Surveys of staple items in the diet* provide a more detailed picture and are a very effective way of assessing the extents and levels of exposure of the majority of the population to pesticides residues in the main items of their diet, as well as any changes which may occur over time. For example Fig. 8.2 summarizes the results of this type of surveillance by the UK Working Party on Pesticide Residues in 1988–9 and 1989–90. It was clear that pesticides residues were detected in the minority of samples, and that residues concentrations were below the MRLs in all but a very few samples.

— *Surveillance of individual foodstuffs* is obviously a labour-intensive way of studying exposure to pesticides residues, but it does provide the most detailed picture. In some cases this is probably unnecessary as the incidences and levels of residues are so low that there is little chance that MRLs or ADIs will be exceeded. However, this type of survey is essential where a potential problem needs to be defined or to provide specific reassurance that there is not a problem with a particular pesticide or foodstuff. For example Table 8.5 summarizes a survey of four pesticides in maize imported into the UK. Previous work had shown that residues could be present, but specific surveillance was carried out to see if MRLs were being exceeded. The survey provided some reassurance because the highest residues levels found in most cases were below the respective MRLs. However, where MRLs are apparently exceeded sampling methodology and the performance of analytical methods must be considered before judging whether the MRL is actually being exceeded (see sections 9.5 and 11.4).

In practice a combination of these approaches is used to define the incidences and levels, and hence consumer intake, of pesticides residues in the diet in the UK and several other western countries. In areas of the world where less resource is available to build up data on pesticides residues in food there is often concentration of effort on the analysis of dietary staples for the more persistent pesticides, particularly organochlorine pesticides. This can be very helpful where these older pesticides are still used or are present at high levels in the environment, but it is of little help if newer, less persistent pesticides replace the older alternatives (although there are factors which might delay this change—section 8.3.2). For this and other reasons several developed countries carry out surveillance for pesticides residues in imported as well as home-produced foodstuffs. Some recent UK

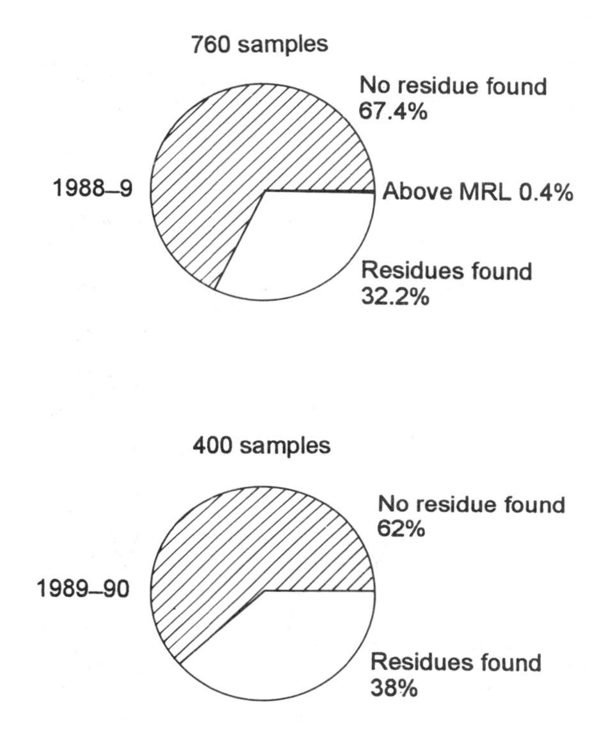

Fig. 8.2. Surveillance of dietary staples. The pie charts summarize this part of the surveillance by the UK Working Party on Pesticide Residues (which reports on the Steering Group on Chemical Aspects of Food Surveillance—see Fig. 9.2). The dietary staples were bread, milk and potato.

Table 8.5. Survey of imported maize for four pesticides

Pesticide	Maximum residue concentration (mg/kg) detected	Maximum Residue Limits (mg/kg)
Malathion	2.3	8
gamma-HCH	0.007	0.5
Dieldrin	0.04	0.02†
Pirimiphos-methyl	0.25	10

† The MRL was exceeded in the case of two samples from the same shipment.

results (Fig. 8.3) illustrate the type of general comparison that can be made, between imported and home-produced foods.

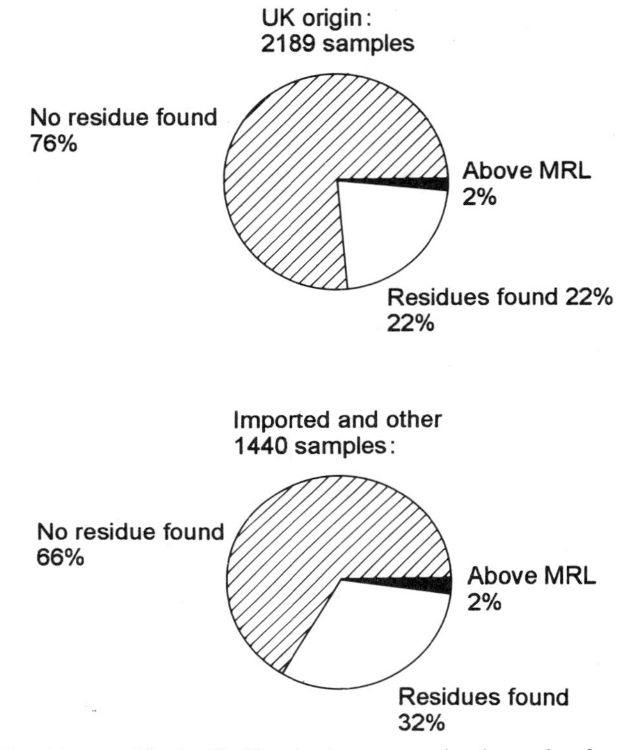

Fig. 8.3. UK and imported foodstuffs. The pie charts summarize the results of surveillance for 1989–90 by the UK Working Party on Pesticide Residues.

Extensive surveillance of pesticides residues in food is carried out in most countries in Western Europe, in the USA, Canada, Australasia and in several of the developing nations in Africa and Asia. For surveillance to be extensive it must cover a very wide range of dietary items and it has to include well over 100 pesticides. Specific surveys of different dietary items are becoming less resource-demanding, as more of the analytical methodology becomes routine, and this has allowed the development of very extensive surveys. For example in a UK survey of fruit and vegetables for pesticides in 1981 to 1984, the crops listed in Table 8.6 were studied. Out of 1649 samples tested, 29 contained residues above the respective MRLs, although several dozen pesticides were included in the surveillance albeit not for each sample. This type of highly-structured study of pesticides residues is still relatively unusual worldwide, but it is a good example of what can be done with the use of considerable resources. The use of extensive resources to survey pesticides residues requires careful management to ensure effective surveillance of the food supply. The same general guidelines apply here as for food surveillance for other chemical contaminants (Chapter 9).

Table 8.6. Commodities analysed for pesticides residues in a UK survey of fruit and vegetables in 1981 to 1984 (The survey was carried out by the UK Working Party on Pesticide Residues.)

Commodity	Analysed for pesticides numbered†
Apples	1 to 12
Apricots	1, 5, 11, 13, 14
Aubergines	1, 10, 12, 15
Avocado pears	1, 5, 11
Bananas	1, 11, 14, 16
Beans	1, 4, 5, 17
Brussels sprouts	1
Cabbages	1, 5, 11, 14, 16
Cane fruit (e.g. blackberries)	1, 14, 16, 17, 18
Carrots	1, 11, 16
Cauliflowers and broccoli	1
Celery	1, 10, 12, 14, 15, 17
Cherries	1, 16
Chinese cabbages	1, 10, 12, 14, 15
Citrus fruit	1, 3, 5, 9, 11, 19, 20
Courgettes	1, 10, 12, 14, 15
Cucumbers	1, 10, 12, 14, 15
Currants	1, 14, 15, 16, 17, 18
Dates	1
Figs	1, 21
Gooseberries	1, 14, 15, 16, 17, 18
Grapes	1, 4, 9, 11, 12, 14, 16, 17
Kiwi fruit	1, 10
Lettuces	1, 10, 12, 14, 15, 17
Marrows	1, 10, 12, 14, 15
Melons	1, 10, 12, 14, 15
Mushrooms	1, 5, 11, 12, 14, 15, 17
Nectarines	1, 5, 11, 13, 14
Onions	1
Peaches	1, 9, 11, 14, 16, 17
Pears	1, 4, 9, 16, 22
Peppers	1, 10, 12, 14, 15
Pineapples	1, 9, 21, 23
Plums	1
Potatoes	11, 23, 24, 25, 26, 27, 28
Radishes	1, 10, 12, 14, 15
Strawberries	1, 14, 15, 16, 17, 18
Tomatoes	1, 10, 12, 14, 15

† Key to pesticides that were analysed in samples of fruit and/or vegetables:
1: organochlorine and/or organophosphorus pesticides; 2: binapacryl; 3: biphenyl; 4: captan; 5: carbendazim; 6: diphenylamine; 7: dithianon; 8: ethoxyquin; 9: 2-phenylphenol; 10: pyrethroids; 11: thiabendazole; 12: vinclozolin; 13: dichloran; 14: iprodione; 15: chlorothalonil; 16: benomyl; 17: dithiocarbamates; 18: dichlofluanid; 19: imazalil; 20: 2,4-D; 21: ethephon; 22: permethrin; 23: captafol; 24: *sec*-butylamine; 25: chlorpropham; 26: dichlorophen; 27: dinoseb; 28: tecnazene.

8.3.5 Research

There has been very considerable scientific research, both to help develop the surveillance of pesticides residues in food and for a wide range of other purposes, not least to provide dossiers about residues to those approving new pesticides. Thus research on pesticides residues covers a diverse range of activities, including the following.

- — Carrying out experimental trials to determine the range of residues levels in treated crops or farm animals.
- — Studying the routes by which pesticides contaminate the food chain.
- — Research on pesticides residues in human milk and fat (e.g. see Table 3.5).
- — Toxicological studies, for example to allow ADIs to be established.
- — Work on the fate of pesticides in the food chain, for example on the chemical modification of pesticides by their metabolism in plants and farm animals. There is also research on the effects of cooking and other forms of processing on pesticides in food.

All of this work requires highly sensitive methods of analysis for pesticides residues, and of course careful experimental design. There is a growing library of information about how to carry out many of the tests used in the above research areas. For example to help toxicological studies on these and other chemicals, the UK Department of Health has issued *Guidelines for the Testing of Chemicals for Mutagenicity* (HMSO, London, 1989; 35th Report on Health and Social Subjects). There is also extensive literature on the review of residues data and related information in the approval of pesticides (see for example the Advisory Committee on Pesticides' Annual Reports noted at the end of this chapter). This literature provides an essential background to surveys of pesticides in food as well as information about how pesticides are assessed and controlled.

8.4 CONCLUSIONS

Study of pesticides residues in food has been carried out for many years, and has produced a very considerable amount of information about their levels and incidences in the food supply, in the UK and many other countries. This information is essential in controlling the residues in food. Sophisticated mechanisms of control have evolved and been applied in the UK, other developed countries and in some Third World countries. International trade in food is subject to similar controls. Effective control depends crucially on finding the source(s) of residues when a problem is identified—a general rule for controlling chemical contaminants in food. It is essential to carry out surveillance of the food supply for pesticides residues—it helps to identify problems, trace back and locate the source of a problem, and check that control has been effectively applied by demonstrating whether the problem has been solved.

Work on pesticides residues in food is largely based on the development of sensitive and reproducible methods of analysis. Without these methods it would not be possible to carry out the research which is essential for the approval and control of pesticides. It is because good methods of analysis have been available for several years that so much is known about many pesticides that can give rise to residues in food. It has also been established that some pesticides are very unlikely to give rise to residues in food. It is important to have extensive data on those pesticides that might be present in food. Such data are available for food and drink in the UK and many other countries.

FURTHER READING

Hassall, K. A. *The chemistry of pesticides: their metabolism, mode of action and uses in crop protection*. Macmillan, London. (1982)

Worthing, C. R. (ed.) *The pesticide manual—a world compendium*, 8th edn. The British Crop Protection Council, Thornton Heath, UK. (1987)

Food Surveillance Paper No.16. HMSO, London. (1986)

Food Surveillance Paper No. 25. HMSO, London. (1989)

Food Surveillance Paper No. 34. HMSO, London. (1992)

Advisory Committee on Pesticides' Annual Reports for 1990 and 1991. HMSO, London, 1991 and 1992.

9

Surveillance of food for chemical contaminants

D. H. Watson, Ministry of Agriculture, Fisheries and Food, R242, Ergon House, c/o Nobel House, 17 Smith Square, London SW1P 3JR, UK.

9.1 INTRODUCTION

Surveillance of food for chemical contaminants is carried out in many countries in the world. The objective is to ensure that the diet is safe. The surveillance is largely by the chemical analysis of samples of food (and drink) from the general supply. Analysis of food for polychlorinated biphenyls (Chapter 3, section 3.3.1 and Table 3.3) is one example of this type of work. But food surveillance involves much more than chemical analysis. A properly managed food surveillance programme has the following components:

— Developing and approving adequately sensitive methods of analysis.
— Identifying where there might be consumer risk from particular chemicals in the diet.
— Designing and carrying out surveys.
— Assessing survey results.
— Checking that where action has been taken it has been effective.
— Managing the overall surveillance programme so that it is balanced and is an effective use of resources.
— Communicating the surveillance results.

This chapter reviews in turn each of these major aspects of surveillance for chemical contaminants in food and drink. It is important, however, to bear in mind that surveillance in itself has little intrinsic value. It is by using its results to identify problems that food surveillance contributes to the control of chemical contaminants in the diet. Thus its effectiveness should be measured in terms of how well resources are used in identifying problems, although the use of food surveillance solely to add to current knowledge about chemical contaminants in food or drink is of some academic value.

Food surveillance provides an important contribution to the control procedures used by Government to ensure the chemical safety of food. Where chemicals have to be approved before usage, such as veterinary drugs or pesticides, surveillance provides data on the incidences and levels of chemical residues in the diet. This gives those approving chemicals such as pesticides, veterinary drugs, and food additives, a picture of what the approval process is leading to in terms of the intakes of these chemicals via the diet. This feedback of information can lead to revisions to the conditions in the approvals for specific products, or the withdrawal of approval in some cases. For chemical contaminants that are in food as a result of more indirect routes, such as industrial chemicals, food surveillance can provide the stimulus to investigate contamination, as well as prompting the use of legal or other controls to halt a supply of contaminated food. Food surveillance is also carried out by non-Government organizations. For example many major food retailers are now carrying out surveillance for residues of pesticides, and in some cases veterinary drugs residues and some other chemical contaminants in food, to ensure that their supplies meet Government controls.

The scientific approaches used in food chemical surveillance are fairly similar wherever the work is carried out. The main differences between studies in various countries are the scale of the work, for example in the numbers of samples analysed in surveys, and in the prioritization of surveys on different chemicals. For example in Western Europe there is considerable emphasis on surveillance for metals and pesticides residues in foods, but in several countries in the region there is also extensive surveillance for some other groups of chemical contaminants in foods, particularly veterinary drug residues. In the UK all of these particular groups of chemicals are routinely included in food surveillance by Government. Indeed there is UK surveillance for all of the groups of chemical contaminants discussed in this book. In contrast, many developing countries do not have the resources to carry out extensive surveillance for food chemical contaminants. For example, their priorities can require surveillance for mycotoxins that are produced in the tropics rather than surveys of industrial chemicals. Nevertheless food chemical surveillance is becoming recognized by many as a worthwhile activity, with the amount of data increasing every year. Indeed several mechanisms have been developed in order to help the international exchange of information. In particular the United Nations' FAO/WHO Codex Committee on Food Additives and Contaminants now provides a forum for communicating food chemical surveillance data, with the purpose of harmonizing international trade via agreed controls on the maximum levels of chemical contaminants in food. Together with other Codex committees, which specifically consider pesticides and veterinary drugs (see Chapter 2, section 2.2.4.1, and Chapter 8, section 8.3.2), this helps to promote international communication between those involved in food chemical surveillance in different countries. Whilst there are inevitably differences in how the results of surveillance are interpreted worldwide, there is certainly a need to ensure food chemical surveillance is carried out to a common high standard. International co-operation helps in this as well as helping to build up a picture of what the problems are with chemical contamination in food. This is particularly important where resources may be limited, trade might be hindered or consumers might be exposed to unsafe levels of chemical contaminants in their diet.

9.2 ANALYSING FOOD FOR CHEMICAL CONTAMINANTS

Most surveillance of food for chemical contaminants is carried out by chemical means.

This is because chemical methods allow the quantification of contaminants with confidence, even at the sometimes very low levels they are present in food. It is now possible to detect 10^{-14} g of some contaminants in 1 g of food, for example in the case of dioxins and furans (see Chapter 3: Table 3.1 gives some examples of detection limits). There is less use of biological methods to detect chemical contaminants in food surveillance, although in some cases very sensitive methods are available, because these assays are usually less-readily quantifiable than chemical ones and for several other reasons:

— biological methods (bioassays) are often slower than chemical ones;
— they are usually less specific than chemical methods;
— bioassays tend to have greater degrees of inherent variability so their results can be very difficult to reproduce;
— assays using biological responses can give false-negative results with contamination not being detected (because the assay's response is less predictable than a chemical one) and false-positive results which are positive results where no contamination is present (for the same reason that false-negative results are given and also because bioassays are often of broader specificity than necessary for the surveillance being carried out and so can detect chemicals that are not contaminants). Whilst both of these effects can also occur with chemical methods, they can be readily eliminated by using suitably sensitive methods with confirmation by another chemical method.

Where chemical methods of analysis have not been developed, and a biological test is available, these shortcomings can be taken into account. For example algal toxins and bacterial toxins are still usually detected using biological assays (Chapter 5). Nevertheless some of the chemical methods now used in surveillance for chemical contaminants in food use techniques that are based on biological interactions, notably affinity chromatography and immunoassay. The first of these techniques relies on a biologically specific interaction, for example between an enzyme and its cofactor, to separate chemical mixtures. The second type of method uses the very strong interaction between antibody and antigen to assay chemicals in food extracts or even in food. The strength of the antibody-antigen reaction makes immunoassay a very sensitive method of detection. The narrow specificity of this reaction makes it a suitable technique for screening unresolved or partially separated mixtures of chemical contaminants and other chemicals in food. Table 9.1 summarizes the principle of immunoassay and provides an example of its use in food surveillance. Immunoassay has the potential for extensive use in food chemical surveillance, as it is quick and relatively straightforward to apply. Affinity chromatography is less widely used in food chemical surveillance, although it has considerable potential as the strength and specificity of the biological interactions which it uses could lead to very considerable purification of food extracts in one step (see Chapter 2, section 2.5.5).

The great majority of chemical methods of analysing food for contaminants use the same general approach: extraction of food samples using solvents, 'clean up' (purification) of the extracts using solvent partition and/or column chromatography, final separation of the chemical(s) of interest, detection of the chemical and confirmation of its presence. A very considerable amount of work has been devoted to the development of different ways of carrying out each of these stages over the last 30 years (see for example Chapter 2, section 2.5). And this developmental work continues even though methods of

Table 9.1. Immunoassays

(a) *Principle*: The assay depends on two competing reactions:

$$Ab + Ag \rightarrow Ab.Ag$$

$$Ab + Ag^{L} \rightarrow Ab.Ag^{L}$$

Antibody (Ab) reacts either with antigen (Ag) or with labelled antigen (Ag^{L}). Competition for available antibody, by Ag^{L} and Ag, allows the amount of Ag to be determined if Ag^{L} and Ab are present in known amounts. In the case of food surveillance for chemicals the contaminant is Ag. The amount of Ag present is determined by mixing it with antibody and a labelled form of the chemical (Ag^{L}; labelled with radioactivity for example). The reduction in detectable Ag^{L} caused by binding of Ag to Ab allows the concentration of Ag to be determined.

(b) *Example*: Immunoassay was used in surveillance for residues of the synthetic stilbene oestrogens in animals in Great Britain in 1980 to 1986. Samples from slaughterhouses were tested by radioimmunoassay, with confirmation of results by another immunoassay or chemically. The following results were obtained:

		Numbers of samples		
Types of animals	Samples tested	Tested	Positive	Levels of stilbenes found
Cattle, calves, pigs, sheep	Bile, kidney, liver, meat	4839	67	0.0004 to 4 parts per 10^6

analysis for quite a few chemical contaminants have become well established. This extensive effort to develop new analytical methods has led to a wide range of techniques that the analyst can use. There are now many different ways of, for example, extracting chemical contaminants from food, which can be modified to provide specific methods for most chemical contaminants. There are also many methods for separating chemical contaminants from purified food extracts. Gas chromatography and high performance liquid chromatography are probably the most widely used methods of separation.

With the availability of many methods to detect chemical contaminants in food, it is clearly essential to have full control over the quality of data. There has been considerable effort in several countries to develop schemes that will provide analytical quality assurance (AQA). This has led, for example, to the definition and agreement of criteria in laboratory analysis for particular types of chemical contaminants in food (Table 3.2 in Chapter 3 provides one example of this). There has also been growth in the collaborative testing of analytical methods, for example under the auspices of the US Association of Official Analytical Chemists, and in helping laboratories develop their expertise, such as via the MAFF Food Analysis Performance Assessment Scheme (FAPAS) in the UK (Table 9.2). It is essential that such approaches are adopted by laboratories contributing data to food chemical surveillance exercises—inaccurate data are clearly unacceptable and there must be proper procedures in place to avoid this.

Table 9.2. The Ministry of Agriculture, Fisheries and Food (MAFF) Food Analysis Performance Assessment Scheme

'MAFF has introduced a scheme for voluntary assessment of the technical performance of food analysis laboratories in (the) U.K. It is known as the Food Analysis Performance Assessment Scheme (FAPAS) ...

Reliable analytical data are required for consumer protection purposes:

— to enforce the provisions of the Food Safety Act, and of the Food and Environment Protection Act and of Regulations under both Acts, the European Communities Act and the Medicines Act;
— to estimate dietary intakes of residues and contaminants as part of risk assessment;
— for the official certification of residues/contaminants in meat and meat products for export; (and)
— for use in self-certification particularly in relation to documentation for food exports to EC countries in the Single Market ... (after) 1992 ...

Four steps are being taken by U.K. analytical laboratories to improve the quality of the data they produce:

— By seeking accreditation with the National Measurement Accreditation Scheme (NAMAS). Analogous schemes exist in some other countries. Accreditation rests on (the) use of well-documented analytical methods, adequately trained staff and proper sample identification procedures.
— By the use of analytical methods which have been properly validated by inter-laboratory trials. Such methods have known analytical characteristics and are reliable and robust in use. This ensures that methods are not unacceptably operator-dependent.
— The utilisation of available certified reference materials which contain known levels of the analyte.
— Participating in Analytical Quality Assurance (AQA) Schemes, in which one or more samples are analysed by each participating laboratory and the results are compared with the true figure. Performance assessment schemes of this sort exist for asbestos fibre counting and for lead in blood. There were none in the food contaminant area until the introduction of FAPAS.

Of the four areas of action, it is the use of reference materials and participation in AQA schemes which are crucial in the assessment of laboratory performance. Essentially the first two are intermediate steps towards achieving a good performance but do not ensure or demonstrate that it is being achieved.'

Source: MAFF Food Science Laboratory, Norwich, Fact Sheet No. 9.

9.3 IDENTIFYING WHICH CHEMICALS TO INCLUDE IN FOOD SURVEILLANCE

This requires one to estimate which chemicals might pose a risk to consumers if they were to be found in the diet. Clearly there are several established types of chemical contaminants that have been found in food or drink—the main groups are discussed in previous chapters. But it is important to bear in mind that there may be other chemicals outside these groups that could be worth including in food surveillance work (this question is discussed in Chapter 11, section 11.2).

The types of questions that need to be asked when considering which chemicals to include in food chemical surveillance work are as follows:

— *What is known about the toxicology of the chemicals?* There is little or no point in carrying out surveillance for non-toxic chemicals in food, except perhaps where there is an environmental objective such as reducing contamination to a level that is as low as reasonably achievable (see Chapter 3, section 3.5). But where the objective is to protect food safety it is essential to review the toxicology of the chemicals first and not to assume that it will develop to a suitable degree by the time the surveillance work has been carried out.

— *What is the evidence that the particular chemicals might be found in food or drink?* For chemicals that have been extensively surveyed in food—pesticides and veterinary drug residues, nitrate, metals, for example—it is essential to research the scientific literature to check what has been found in *relevant* foods or drinks. Relevance in this context is defined by the objectives of the surveillance. For example work on national food supplies might draw on the results of food surveillance in other countries with similar dietary habits. For the less extensively studied chemicals in food, such as dioxins, this is clearly not likely to provide sufficient information. In this circumstance it might be possible to draw on the results of environmental surveillance—to provide some clues as to the best types of foods and the particular chemicals that are most worthwhile concentrating on in food surveillance. However, the key point for all types of chemical contaminants is to draw on data and knowledge from all relevant sources in reviewing what surveillance might be done. Where there is no information at all, but there is still suspicion that the diet is contaminated with a chemical, particularly where the chemical is very toxic (such as benzene, Chapter 3, section 3.5), it is essential to carry out laboratory research to provide information from which one can judge whether food surveillance is likely to be worthwhile. It is possible to help this by developing information systems for identifying high priority chemicals for food surveillance (e.g. the prioritization scheme described in Chapter 3, section 3.1). But this type of approach is likely to work best where there is information about the likely levels and incidences of contamination by the chemical in food. This usually requires some practical research where there are few data in the scientific literature.

— *What methods of analysis are available?* This is a secondary consideration compared with the above questions because if it is judged that a chemical is sufficiently hazardous and that it could be present in food, these should be sufficient grounds to develop suitable methods of analysis. Nevertheless where finite resources can only allow surveillance for a set number of chemicals, and questions of toxicology or likely occurrence in food give equal priority to many chemicals, surveillance of those substances for which methods of analysis are available could be one way of prioritization.

These questions should provide a reasonable basis on which to identify priority chemicals for food surveillance work. But there should always be room for developing priorities as the surveillance work proceeds. The value of initial research before embarking on food surveillance cannot be overemphasized, as sound information is essential if surveillance is to identify where the problems are. Although not finding a chemical contaminant in a survey can be reassuring, it is also a waste of resources if problems go undetected for lack of initial research.

9.4 DESIGNING AND CARRYING OUT SURVEYS OF FOOD FOR CHEMICAL CONTAMINANTS

Table 9.3 lists the main points that need to be considered, together with some comments on how the main questions can be answered. There are few guidelines and no firm rules about how food chemical surveillance should be carried out. The comments in this table derive from personal experience of carrying out such work. There are many different views on how best to address the questions in Table 9.3. The best way of seeing whether or not a survey is effective is to compare its results with those of similar surveys. Where this is practicable it is surprising how often different surveys of a given chemical contaminant provide similar pictures. Hence the parameters noted in Table 9.3 might not be as important as those people involved in food chemical surveillance might sometimes think. In the absence of scientific work on the best ways of obtaining representative samples (see Chapter 11, section 11.3), it can be difficult to tell how important the questions in Table 9.3 are.

Although there is some doubt about the best ways of designing surveys for chemical contaminants in food, it is still essential to carry out a survey in the most effective way possible. Some general comments on how this can be done are as follows.

— It is important to develop an integrated but flexible plan for the supply and analysis of food samples well before surveillance starts. Unforeseen problems occur, usually with obtaining the required number of samples, or the right type of sample. A rigid quota of samples of a particular brand of foodstuff is likely to be unachievable. It is realistic to provide a choice of brand names and a flexible quota of samples, with firm lower and upper limits on the required numbers of samples.

— Communication must be effective. Good and regular communication between the person organizing the work, those obtaining samples, and the analysts is a key factor in carrying out effective surveillance of the food supply. This is particularly important when the survey is starting, to ensure that it is developing well and that 'teething problems' are dealt with. It is also important as the survey starts to provide feedback of results to all those involved. This helps to motivate those obtaining samples as well as the people managing the survey or carrying out analytical work.

— The survey's objectives must be clear. The objectives of a survey may need to be refined or changed as the surveillance progresses, but the objectives, new or old, must be clear to all concerned if the survey is not to lose its way.

Table 9.3. Main questions involved in designing food chemical surveillance exercises

Question	Comment
How many food samples need to be analysed for the contaminant?	There is usually no statistical basis for deciding this in food chemical surveillance (but see Chapter 2, section 2.4.3). As a 'rule of thumb' less than 20 samples is likely to give too narrow a range of results even if all the samples contain the contaminant. More than 100 samples per chemical contaminant can be wasteful of analytical resources, but where the incidence of contamination is very low many more than 100 samples might need to be analysed (e.g. see Table 9.1(b)).
Where and when should the samples be obtained?	This depends on the structure of the food supply network, and the need to have a sufficiently representative set of samples from enough locations and during the relevant times of the year. If contamination is likely to be seasonal or localized by region, structure the survey accordingly.
How sensitive should the analytical method be?	Where there is a toxicologically-based limit or other form of target, this should be used to define the required limit of detection. Otherwise it is not possible to define a preferred detection limit for the analytical method.

— The scientific quality of the surveillance should be monitored throughout the work. For example, if the detection limit of the analytical method is not quite good enough at the start of the survey, this should be identified and dealt with then, and not later on in the work. Similarly sample quality needs to be carefully controlled—the arrival of twenty batches of thawed, unanalysable samples is twenty batches too many!

— An agreed time limit should apply. It is sometimes difficult to define the time required to complete a survey until some samples have been analysed, for example if a new analytical method is being used. But then—if not before—a time limit for completing the survey should be agreed.

9.5 ASSESSING SURVEY RESULTS

Again there are several questions to consider. The first point to check is the difference between the survey's structure and its design. The points noted in Table 9.3 need to be addressed again once the survey has been completed.

— Were enough food samples analysed? A significant shortfall in numbers, below the target decided before the survey started, means the survey still needs to be completed.
— Were the samples obtained from the right places and at the correct times?
— Was the analytical method sufficiently sensitive?

At this stage any shortage of adequate data must mean an extension to the survey. Assuming this is not required the survey's results should be reviewed in the context of toxicological information to see if there is an unacceptable risk. How this is done depends very much on the current state of toxicological information about the chemical contaminant. There are several possible ways of starting off this review process, for example:

— *Where there is a defined Maximum Residue Limit (MRL)* it can be a relatively straightforward process to compare the contaminant concentrations found in the survey with the MRL. However, this must take account of the variance in the analytical result. Thus if quantitative results are subject to a ± 20% margin as a result of variation in performance of the analytical methodology, and perhaps sampling problems, any results at less than 20% above the MRL cannot be considered to indicate contaminant at levels above the MRL. In practice analysts generally prefer a larger margin than 20% before contaminant levels are accepted as exceeding the MRL. This is largely due to variations in analytical methodology for the very low levels at which chemical contaminants are usually found in food. It is also important to take account of the non-Gaussian distributions of chemical contaminants in food (Chapter 11, section 11.3.1) which can lead to a few very high results for an otherwise weakly contaminated set of samples.
— *Where there is a toxicological measure of maximum safe intake of the contaminant* (i.e. an Acceptable Daily Intake—ADI—or a Tolerable Daily Intake—TDI) the surveillance data should be used to estimate the intake of the chemical contaminant and this intake estimate be compared with the ADI or TDI. (The methodology used is discussed in Chapter 10.) The use of ADIs or preferably TDIs is becoming common for chemical contaminants in food. The use of ADIs for food additives is well established, and they have also been used quite widely for pesticide residues, mainly as a result of reviews by the United Nations' WHO Joint Meeting on Pesticide Residues. The comparable committee for other chemical contaminants is the Joint FAO/WHO Expert Committee on Food Additives, which despite its title also reviews the toxicology of chemical contaminants, including metals, natural toxicants and veterinary drugs residues. TDIs for chemical contaminants are also defined by national expert bodies such as the Committee on Toxicity of Chemicals in Food, Consumer Products and the Environment in the UK, and by supranational committees such as the EC Commission's Scientific Committee for Food. It is essential to consult toxicologists to take account of any variations in the TDIs that have been recommended by several

committees. However, in some cases the same TDI is proposed by the committees. Where there are differences they are usually small.

— *Where there are no TDIs, ADIs or MRLs* it is essential to determine why this is the case before reviewing the survey data. For some carcinogenic chemicals it is not possible to set a TDI, ADI or MRL because a cancerous effect can be observed when even very low levels of the chemical are ingested by experimental animals. Where there is not an identifiable threshold for a toxicological effect (i.e. it is non-stochastic), the presence of the chemical contaminant in the diet is generally enough to indicate a potential problem. Where a toxicological threshold has not been identified, but the chemical contaminant is suspected to have one, it can obviously be very difficult to identify the extent of risk from survey results. Sometimes this stimulates further toxicological work. Alternatively the ALARA approach (Chapter 3, section 3.5) can be brought into play. However, this is unusual for non-carcinogenic chemical contaminants in food, especially where further experimental work on the chemical's toxicology is needed. In this situation it is imperative that further toxicological work is done so that the survey results can be interpreted fully. The pitfalls of not following this type of approach are clear in the case of many mycotoxins (Chapter 5, section 5.4), where the significance of much chemical investigation work is difficult to determine because toxicological experiments need to be carried out.

Tables 9.4, 9.5 and 9.6 illustrate some of the above points using data from two UK surveys for pesticides residues. Several pesticides were detected in dietary samples although many more pesticides were not found. Table 9.4 gives some data for one of the detected pesticides. The various 'food groups' are mainly composites of different types of foodstuffs which are purchased in the UK on a regular basis. The survey data are relatively meaningless on their own, but if they are converted to a figure for dietary intake this provides some meaning. In this case the survey data were converted to an intake figure by multiplying them with values for the computed average intakes of the food groups.

Table 9.4. γ-HCH residues (mg/kg) in the UK Total Diet Study

Food group	γ-HCH levels
'Other cereal products'	ND[†] to 0.002 (< 0.0005)
Carcase meat	ND to 0.09 (0.006)
Offals	ND to 0.02 (0.0005)
Poultry	ND to 0.08 (0.005)

Table 9.4. (continued)

Food group	γ-HCH levels
Fish	ND
Eggs	ND to 0.004 (< 0.0005)
Potatoes	ND to 0.005 (< 0.0005)
Milk	ND
Dairy products	ND to 0.02 (0.001)
Nuts	ND to 0.009 (0.0005)

† ND = Not detected; mean concentrations are given in parentheses.

Table 9.5. Computed average intakes and ADIs and for some pesticide residues detected in the UK Total Diet Study in 1984–5

	Intakes (mg per 70 kg person per day)	
Pesticide	Computed average intakes	Acceptable Daily Intakes
γ-HCH	0.0005	0.7
Chlorpyrifos	0.0001	0.7
Malathion	0.0001	1.4
Pirimiphos-methyl	0.0018	0.7
Cypermethrin	< 0.0001	3.5
Permethrin	0.0001	3.5
Quintozene	0.0002	0.49
Tecnazene	0.0043	0.7

Intakes were computed using residues data, including those for *g* -HCH in Table 9.4, and standard conversion factors to take account of the amount of different types of food that are consumed.

Adding up the resulting figures in Table 9.4 for the various food groups gave an estimate of average intake for the contaminant. The estimated average intake for γ-HCH and the other chemicals detected in the survey can then be compared (Table 9.5) with the ADIs defined by an authoritative body, in this case the Joint FAO/WHO Meeting on Pesticide Residues. It is also possible to compare the residues concentrations of many pesticides with MRLs. Table 9.6 gives an example of this in which residues of 'total

DDT' found in a survey of fruit and vegetables in 1985 to 1987 are compared with a European Community MRL. These types of comparisons are quite easy to make, although the methodology of estimating intakes can be quite complex and it is essential not to overinterpret intake estimates (Chapter 10).

Table 9.6. Comparison of survey results with an MRL for 'Total DDT'

'Total DDT' = sum of *pp′*-DDT, *op′*-DDT, *pp′*-DDE and *pp′*-TDE. Surveillance was carried out in Britain using retail samples in 1985–7.

	Numbers of samples			
Samples	Tested	Containing residues	Containing residues above the MRL	Residues concentrations (mg/kg) above the EC MRL
Strawberries	236	11	1	0.2
Brussels sprouts	302	9	5	0.2–0.7
Cabbage	323	9	6	0.2–2.4

9.6 USING FOOD SURVEILLANCE TO CHECK THAT PROBLEMS HAVE BEEN SOLVED

Much of the above-described approach to food surveillance is about identifying problems. This should, of course, be followed by action to reduce exposure to the chemical contaminants. A discussion of the variety of available control methods is beyond the scope of this chapter (but see Chapters 2 and 8 for some examples). It is worth emphasizing that the means adopted to control chemical contamination of food should not dictate how follow-up surveillance should be carried out once control has been given a chance to take effect. It is important to carry out subsequent surveillance to check that control has been effective or, if it has not, to stimulate further action. An example of a 'follow-up' survey is given in Fig. 5.2 where initial surveillance demonstrated a high incidence of aflatoxin M_1-contaminated milk samples. After action was taken another survey showed that the control measures had the effect of reducing the incidence of contaminated milk samples by a dramatic margin. The change was unusually quick—probably because one single source of contamination was involved and it was therefore possible to halt the contaminant flow at source. More often there is a steady decline in contaminant levels and/or the incidence of contaminated samples. Fig. 9.1 illustrates the type of change that is more usually detected. Clearly such changes can only be detected over a long period of time.

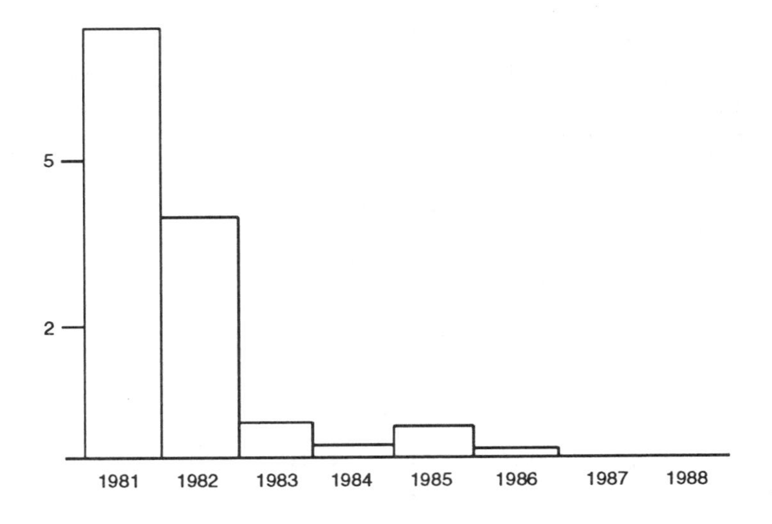

Fig. 9.1. Decline in the incidence of stilbene residues in food-producing animals in Britain. The use of stilbene growth-promoting hormones in food-producing animals such as cattle, calves, pigs and sheep, was prohibited in 1982. Those found guilty of their illegal use have been fined. As the graph shows, a marked decrease in the incidence of residues has been measured.

9.7 MANAGING FOOD CHEMICAL SURVEILLANCE PROGRAMMES

The above sections describe much of the detail necessary to carry out individual surveys. Surveillance needs to be co-ordinated and otherwise managed as a programme, to ensure that resources are targeted effectively. There are several different ways of doing this. In the UK, Government draws on the advice of independent experts. A committee, the Steering Group on Chemical Aspects of Food Surveillance (SGCAFS), advises the Government on their programme of surveillance for chemical contaminants (and nutrients and additives) in food. Work on the full range of chemical contaminants is reviewed regularly. Detailed review of work on each group of chemicals, pesticides residues, natural toxicants and so on, is provided by expert working parties reporting to the SGCAFS. The structure of this system of committees includes inputs from other groups of independent experts such as the Advisory Committee on Pesticides, and the Veterinary Products Committee (Fig. 9.2). In other countries the food chemical surveillance programmes carried out by governments do not necessarily draw on such a formalized system of peer review and advice, although most national programmes in western Europe and north America rely to some degree on independent advice. It is important to have some element of peer review in managing a food chemical surveillance programme since the work must be carried out to the highest scientific standards. Indeed the methods and results are open to challenge when they are published so there is no reason why this should not also be the case whilst the work is going on.

Some of the main factors to be considered in managing food chemical surveillance programmes are as follows:

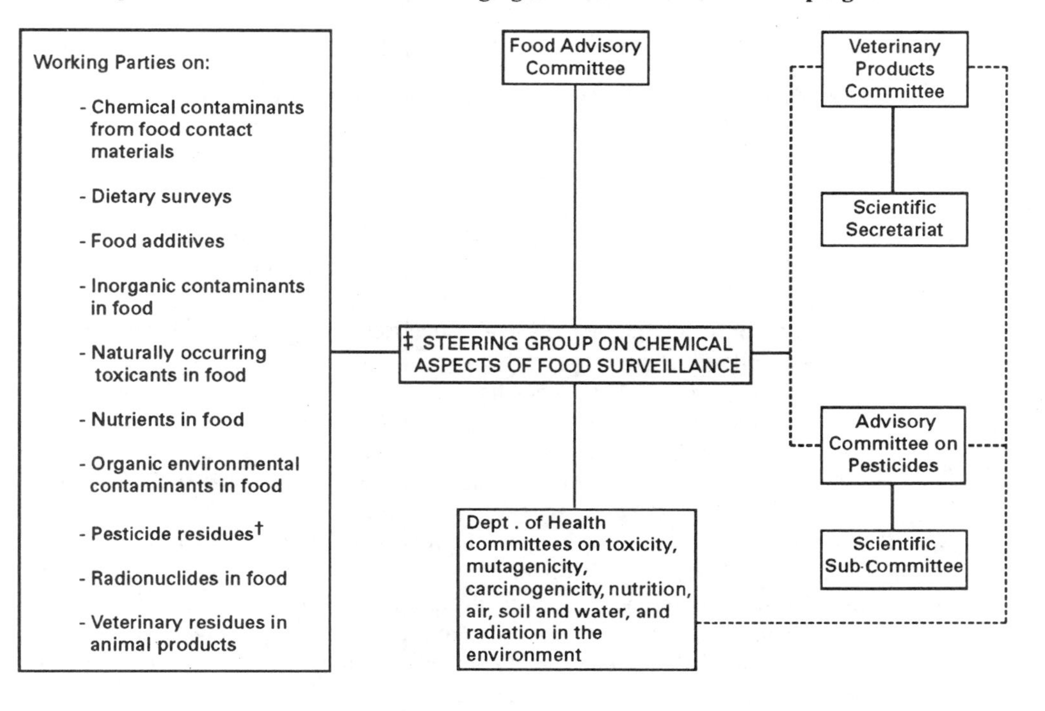

Fig. 9.2. The UK Steering Group on Chemical Aspects of Food Surveillance in relation to other committees. Solid lines between committees indicate formal links whilst broken lines show informal connections.

— *Is there unnecessary duplication of effort?* It is important, for example, to make the best use of food samples. Co-ordination of surveys can sometimes avoid unnecessary duplication of effort in obtaining samples. And the same samples can be analysed for several different chemical contaminants.
— *Are the objectives of the survey clear and well thought out?*
— *Are costs being controlled effectively?* It is difficult to provide a unit cost for analysing a sample of food for a variety of different chemical contaminants, but it is possible to compare costs of analysis (and sampling) where the same chemical is involved.
— *Is enough attention given to achieving a high quality of surveillance data?* This can, of course, push up costs but it needs to be considered for a programme of surveys, together with the objectives and the costs of the work.
— *Has action, that was taken after surveillance found a problem, been effective?* This involves considerable judgement, since action is not always dramatically or immediately effective (section 9.6).

9.8 COMMUNICATING SURVEILLANCE RESULTS

Communicating the results of surveillance work is a key factor in the management process. The work should be reported not only to a peer review system such as that in Fig. 9.2 but also to the consumer, scientists and others outside an organization. The main external communication routes are as follows:

— *Government reports* In the UK, *Food Surveillance Papers* are the primary means used to report all of the Government's food chemical surveillance work (Table 9.7 lists the reports produced so far). It is important to look for the toxicological and other types of interpretation of surveillance results in such reports (this is provided in appendices to Food Surveillance Papers).

— *Scientific journals* There are a few specialist journals that include several articles per issue on food chemical surveillance work, notably *Food Additives and Contaminants* (UK) and *Vår Föda* (Sweden). Other journals which less frequently contain papers on food surveillance include *Journal of the Science of Food and Agriculture* (UK), *Journal of Agricultural and Food Chemistry* (USA), *Journal of the Association of Official Analytical Chemists* (USA), *Archiv für Lebensmittelhygiene* (Germany), and *Alimentaire* (France). Because of the diversity of food science journals, most of which publish food chemical surveillance work from time to time, it is essential to scan the literature widely to build up an effective database. Even so computerized on-line searches are usually essential when a survey is being considered or designed, to ensure that all relevant information is taken into account.

— *Symposia* can be an excellent source of up-to-date information, particularly about methods of analysis. There are fewer symposia on the results of food chemical surveillance.

— *International organizations* For example the United Nations' Global Environment Monitoring System (GEMS) collates much food chemical surveillance data from governments. The resulting GEMS/Food reports provide a more general picture of chemical contaminants in food and the environment than many other sources. This work concentrates on residues in food of pesticides, PCBs, aflatoxins, lead, cadmium, mercury and tin.

Table 9.7. Food Surveillance Papers

Food Surveillance Paper No.	Title
1	The surveillance of food contamination in the United Kingdom
2	Survey of vinyl chloride content of polyvinyl chloride for food contact and of foods
3	Survey of vinylidene chloride levels in food contact materials and in foods
4	Survey of mycotoxins in the United Kingdom
5	Survey of copper and zinc in food

Table 9.7. (continued)

Food Surveillance Paper No.	Title
6	Survey of acrylonitrile and methacrylonitrile levels in food contact materials and in foods
7	Survey of dieldrin residues in food
8	Survey of arsenic in food
9	Report of the Working Party on Pesticide Residues (1977–1981)
10	Survey of lead in food: second supplementary report
11	Survey of styrene levels in food contact materials and in foods
12	Survey of cadmium in food: first supplementary report
13	Polychlorinated biphenyl (PCB) residues in food and human tissues
14	Steering Group on Food Surveillance progress report 1984
15	Survey of aluminium, antimony, chromium, cobalt, indium, nickel, thallium and tin in food
16	Report of the Working Party on Pesticide Residues (1982 to 1985)
17	Survey of mercury in food: second supplementary report
18	Mycotoxins
19	Survey of colour usage in food
20	Nitrate, nitrite and *N*-nitroso compounds in food
21	Survey of plasticiser levels in food contact materials and in foods
22	Anabolic, anthelmintic and antimicrobial agents
23	The British diet: finding the facts
24	Food surveillance 1985 to 1988
25	Report of the Working Party on Pesticides Residue: 1985–88
26	Migration of substances from food contact materials into food
27	Lead in food: progress report
28	Programmes to monitor radioactivity in food
29	Intakes of intense and bulk sweeteners in the UK 1987–1988
30	Plasticisers: continuing surveillance
31	Dioxins in food
32	Nitrate, nitrite and *N*-nitroso compounds in food: progress report
33	Veterinary residues in animal products 1986 to 1990
34	Report of the Working Party on Pesticide Residues: 1988–90
35	Food chemical surveillance 1989 to 1992
36	Mycotoxins: third report

These reports are available from HMSO bookshops and their accredited agents (in the UK and overseas) or can be ordered direct by telephone (071-873-9090 in the UK or 01044-71-873-9090 from other countries).

In considering the results reported via these routes it is important to ask the type of key questions about the work that have been discussed in this chapter. The quality of food chemical surveillance, and hence the results, can vary quite considerably from chemical to chemical, as well as from organization to organization. A critical review of the work needs to be done by the reader every time he or she works through a report or article, or listens to lectures at a symposium. It is hoped that this chapter provides some guidance on how this can be done.

FURTHER READING

The reader is recommended to review the various journals and reports noted above in developing his or her approach to food chemical surveillance work. Some more general references on the methodology involved in food chemical surveillance are as follows:

Bell, J. R. and Watson, D. H. Food production contaminants: control and surveillance. In: *Food contaminants: sources and surveillance*. RSC, Cambridge, UK, pp. 61–72. (1991)

Food Surveillance Papers (listed in Table 9.7).

Food monitoring in Denmark: nutrients and contaminants 1983–1987. Levnedsmiddelstyrelsen, Copenhagen. (1990)

Knowles, M. E., Bell, J. R., Norman, J. A. and Watson, D. H. *Food Additives and Contaminants*, **8**, 551–64. (1991)

Gordon, M. H. (ed.) *Principles and applications of gas chromatography in food analysis.* Ellis Horwood. (1990)

10

Estimating consumer intakes of food chemical contaminants

N. M. A. Rees and D. R. Tennant, Ministry of Agriculture, Fisheries and Food, Rooms 239B and 235 respectively, Ergon House, c/o Nobel House, 17 Smith Square, London SW1P 3JR, UK.

10.1 INTRODUCTION

Previous chapters have discussed the structures and properties of some food chemical contaminants and the planning and results of food surveillance programmes. Of course the aims of food surveillance programmes are many and various. Among these aims is the assemblage of data so that the intakes of food chemicals can be estimated. By intakes, in this context, we mean the amounts of food chemicals that are present in the food which consumers actually eat. So far this book has described ways of determining the incidences and concentrations of chemicals in food. In order to estimate consumers' intakes some further important information is required: the amounts eaten of the foods in which the chemical occurs. This information is obtained from a different kind of food surveillance—dietary surveys, which are designed to gather information on the types and amounts of food in people's diets. The purpose of this chapter is to discuss the various methods available to obtain information on people's diets and to estimate intakes of chemical contaminants from food.

Total diet studies (TDS) and duplicate diet studies (DDS) have also been mentioned previously as methods for measuring consumer intakes of chemical contaminants. They will be discussed with the other available methods later in this chapter (section 10.4).

Estimating the intake of contaminants presents unique problems for two main reasons. Firstly, contaminant concentrations in foods cannot be predicted easily and are influenced by many variable factors, such as geological and atmospheric deposits, local industry, farming practices, and food storage and preparation. The range of concentrations can therefore be very wide even in foods grown under apparently the same conditions. Secondly, some contaminants are found in air and water as well as in food and their contribution also needs to be allowed for if the total human intake is to be assessed. For example, areas of urbanization may have higher concentrations of aerial pollution, and rural areas may have groundwater with high levels of metals because it runs through old mine workings.

10.2 RISK ASSESSMENT OF FOOD CONTAMINANTS

Estimates of intakes of food chemicals are usually undertaken in order to assess the potential risk from those chemicals in the diet. A three-stranded approach is employed in the risk assessment of food chemicals, although the terminology may vary from method to method (Fig. 10.1).

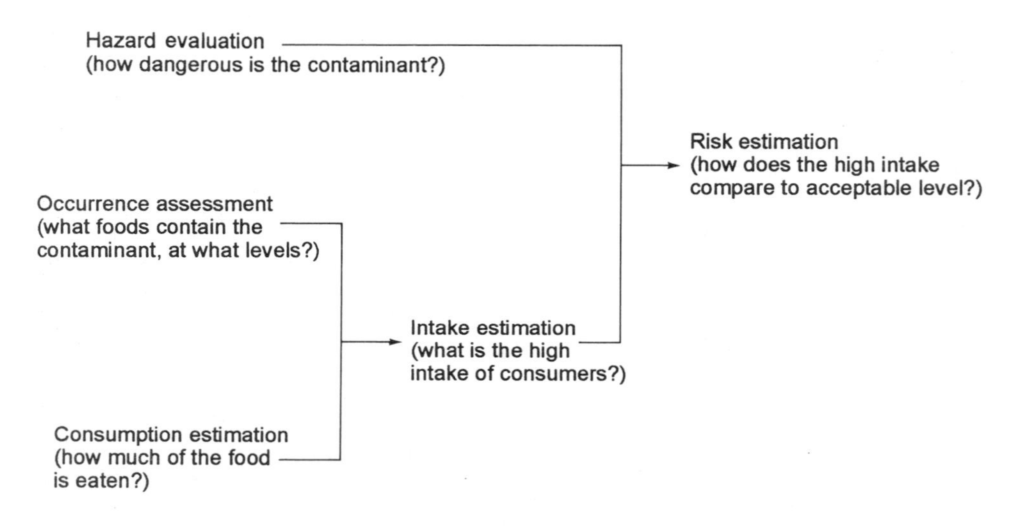

Fig. 10.1. General scheme for food chemical risk assessment.

The three essential elements of the risk assessment process are: the *hazard evaluation*, where all the available toxicological or epidemiological data are considered, the *occurrence assessment*, which seeks to establish the concentration or incidence of the hazard of concern in foods; and the *consumption estimate*, which seeks to establish how much of the foods (called source foods) of concern consumers actually eat. In other words: how hazardous is the food chemical? what foods contain it? and how much of those foods are eaten?

10.2.1 Evaluating hazard

The UK Department of Health's Committee on Toxicity of Chemicals in Food, Consumer Products and the Environment (COT) (Fig. 9.2) provides advice to the Government on the toxicity and on the possible health hazards from contaminants in food.

Since 1989 the COT has provided its advice in terms of acceptable daily intakes (ADIs) for chemicals added to food, and tolerable daily intakes (TDIs) for chemical contaminants (see Glossary). It should be emphasized that the term 'tolerable' implies permissibility rather than acceptability. It is used where intake of a contaminant isunavoidably associated with consumption of otherwise wholesome and nutritious food.

The tolerable level can be expressed as a daily intake (TDI) or weekly intake (TWI) or monthly (TMI). These intakes are prefixed with the letter P (e.g. PTWI) if they are provisional.

The TDIs (or TWI, TMI etc) are based on the highest intake (expressed as mg/kg bodyweight per day) which gives rise to no observable adverse effects, usually in animals. Appropriate safety factors (usually 100) are then applied to allow for interspecies extrapolation and inter-individual variation. The acceptable daily intake (ADI) is set in a similar fashion for pesticides and veterinary drug residues.

10.2.2 Estimating occurrence

The concentrations and incidences of chemicals in food can be determined using the methods of food surveillance described in Chapter 9. However, surveillance data may not always be available or necessary for risk assessment purposes. For example, it may be necessary to predict the likely level of intake if a particular regulatory limit for a contaminant in foods were to be set. In this case the proposed limit value could be used in conjunction with data on the consumption of foods to calculate the intake. Of course care would need to be taken in the interpretation of such a model since it is unlikely that the contaminant would occur at the regulatory limit in all foods all of the time. More discussion of the use of different sources of occurrence data is given in section 10.3.3.

10.2.3 Estimating consumption

Estimating the consumption of food is also not a simple matter. Various methods are available and these will be discussed in more detail later in this chapter in section 10.4. However, three broad approaches are available: *total diet (market basket) studies* where people's expenditure on food is monitored; *diary surveys* where people are asked to keep a record of all the food they eat; and *duplicate diet studies* where people are asked to prepare identical portions of the food they eat for analysis later.

Depending on the contaminant, additional intakes from water and air may need to be included to give the total estimated intake. Whilst the amounts of air and water used by the 'average' person remain fairly constant within specified age ranges, the pattern of food consumption varies widely (and will be discussed at greater length throughout this chapter).

'Average' water consumption

A WRC (Water Research Centre) survey has provided information on the amount of drinking water consumed by people in various age groups in Great Britain. The results are given in Table 10.1. The average adult consumes about 1 l/day, but consumption may be as high as 2.7 l/day. The amount consumed includes all tap water, whether drunk at home or elsewhere, and beverages made from tap water. It does not include commercial drinks, for example soft drinks or water used in cooking. Infants fed made-up feeds drink approximately 0.15 l/kg bodyweight per day and may take more than this in hot weather.

'Average' air inhalation

The quantity of air inhaled is usually taken as 6 m^3 per day for children and 15 m^3 per day for adults.

Table 10.1. Total consumption of tap water at home and elsewhere in Great Britain

	Total consumption of tap water (l/person per day)		
Age (years)	Minimum	Mean	Maximum
1–3	†	0.50	†
3–14	0.04	0.57	1.54
15+	0.21	1.07	2.73

† Sample size too small for reliable estimate.

'Average' food consumption

Unlike air and water, the quantity of food consumed by the average person is very difficult to determine accurately. The Joint UNEP/FAO/WHO Food Contamination Monitoring Programme (GEMS/Food) has considered the most recent FAO data to estimate the average quantity of food consumed per person worldwide. 'Cultural' diets were developed: African, cereal-based; African, root and tuber-based; North African; Central American; South American; Chinese; Far Eastern; Mediterranean; and European. The total average consumption varied from 862 g/person per day to 1669 g/person per day. This fairly simple study showed that even 'average' diets can vary by large factors.

In the UK the National Food Survey has been conducted on a yearly basis for over 50 years. Some 7000–8000 different households are involved each year, each taking part for one week. The household is requested to keep a record for seven days of the description, quantity and cost of all food and beverage items entering the home for human consumption. Over 200 separate foods are distinguished and averages of the quantities consumed *within a household* are calculated. The average consumption over the last 50 years ranges from 9000 to 11000 g/person per week or 1286 to 1571 g/person per day. However, since the consumption is estimated by dividing the household consumption by the number of people in the household, it may be too high for some and too low for others. This survey has just recently started to include those food and beverages consumed outside the home.

A better estimate of average food consumption by the *individual* can be extracted from the UK Dietary and Nutritional Survey of British Adults (Gregory *et al.* 1990). In this survey the average adult consumed about 2.5 kg of food and beverages a day, but recorded consumption can be as high as 9.7 kg/day or as low as 0.49 kg/day (Table 10.2). The average adult consumes about 1.24 kg of beverages a day and 0.92 kg of food a day (Table 10.2). The amounts given in Table 10.2 include all food or beverages whether eaten at home or outside the home. An infant or child would obviously consume less than these amounts. However, if consumption is expressed on a bodyweight basis infants and children can sometimes consume more per unit of bodyweight than adults.

Table 10.2. Consumption of food and beverages by adults (aged 16–64+) in the UK

	Consumption of food and beverages (kg/person per day)		
Food	Minimum	Mean	Maximum
Solid food	0.10	0.92	2.15
Beverages†	0.45	1.24	9.02
Food and beverages	0.49	2.50	9.74

† Includes tea, coffee, other hot drinks, soft drinks and squashes, milk, etc.

10.2.4 Assessing intake

The occurrence assessment and consumption estimate are combined to estimate the intake of the food chemical by consumers of the foods in question. In some methods data on occurrence and consumption can be obtained simultaneously (section 10.4). If air and water must also be considered as sources of the contaminant, their contributions are added at this stage.

10.2.5 Estimating risk

The intake estimate and the hazard assessment are then brought together to estimate risk by comparing the intake estimate with the tolerable intake. In order adequately to protect the entire consuming population it is not sufficient only to look at the intakes of average consumers. In the UK the Food Advisory Committee has taken the view that the intakes of consumers at the upper end of normal dietary behaviour should be considered. The intakes of these high level consumers are compared to the safety standards set by the COT (i.e. the TDIs for contaminants; section 10.2.1) or by other expert committees. This process is called risk estimation and determines whether the levels of contamination are tolerable, or the levels of pesticides or veterinary drug residues are acceptable. In the simplest case, if an intake is below the tolerable or acceptable intake then no further action is required beyond further surveillance to ensure that the situation does not change. If the estimated intake is above the tolerable intake then some risk management action is required to limit exposure to the food chemical.

10.2.6 Food contaminant risk management

Risk can be managed by addressing either of the factors which contribute to the intake: occurrence or consumption. In other words a risk management exercise could address the following questions:

— Can we eliminate the contaminant?
— Can we decrease the number of foods which contain the contaminant?
— Can we decrease the contaminant levels found in those foods?
— Can we decrease the quantities of those foods which are eaten?

If high intake is indicated in a small defined sub-group of the population then advice on dietary habits, or targeted regulatory action, may be sufficient to manage any potential risk. If high intake is possible in a larger part of the population then it may be necessary to prohibit the food from contaminated sources from entering the food supply. As this would clearly affect people's livelihoods, it is usually based on irrefutable evidence of risk to the consumer or as a temporary precaution until further assessment is conducted. Most intake assessments tend to be conservative so further monitoring, or reassessment of the data, may be indicated if a more accurate assessment is required.

10.3 GUIDELINES FOR ESTIMATING CHEMICAL INTAKES

A great deal of information is required to conduct a risk estimation. Since data on the concentrations of chemicals in food and/or the consumption of the foods under consideration may not always be complete, some guiding principles have been drawn up to be followed when making intake estimates. These guiding principles should be followed regardless of the type of contaminant. These are as follows:

— first and foremost, the way the estimate is calculated should be appropriate for the purpose to which it will be put;
— an assessment of the accuracy of the estimate should be made;
— the assumptions underlying the estimate should be acknowledged and recorded; and
— all estimates must consider possible critical groups.

10.3.1 Appropriateness of dietary intake estimates

There are two important issues to consider:

Firstly what dietary information is available and, of this, which is most suitable to make the estimate? The toxicological properties of the chemical will indicate the survey from which consumption data should be extracted. For example if the contaminant could potentially affect the development of the foetus then the consumption of source foods by women of child-bearing age is the most appropriate to assess the dietary intake. If the contaminant is found in tinned foods commonly consumed by children, for example baked beans, then a specific dietary study on that population would be indicated. It is possible that the higher consumers of the foods of concern are dispersed throughout the population and a large representative study would be a more appropriate source of dietary information.

Secondly, how should the consumption data be expressed? Chemical concentrations or food consumption may need to be expressed in different ways depending on the situation. If a chemical has cumulative or chronic toxic effects, consumption is expressed as an average over a week. It is unlikely there will be sufficient data to estimate the consumption of a particular food over a lifetime, so the weekly average is usually taken as an estimate of the 'lifetime' average. However, the consumption value can be very different depending on the way it is extracted. For example, the average British adult consumes approximately 8 g of orange per day. However, for the 33% of the population that actually consumes oranges in a given week, the average amount consumed is 24 g (or a small orange every five days). The majority of consumers (97.5%) eat 106 g (approximately a small orange every day in an average week) or less. This leaves 2.5% of the consumers who eat more than 106 g in an average day. The maximum consumption (averaged over a week) is 181 g (about a medium-sized orange every day in an average week) which is more than twenty times the average adult consumption.

If a chemical has acute effects the amount which could be eaten in one meal may be more important than the average amount eaten over a given period of time. Fig. 10.2 shows the distribution of maximum 'eating occasion' consumption of oranges by the typical British adult. In comparison to the above example, the maximum recorded consumption by an adult on any one eating occasion is 300 g (one and a half large oranges).

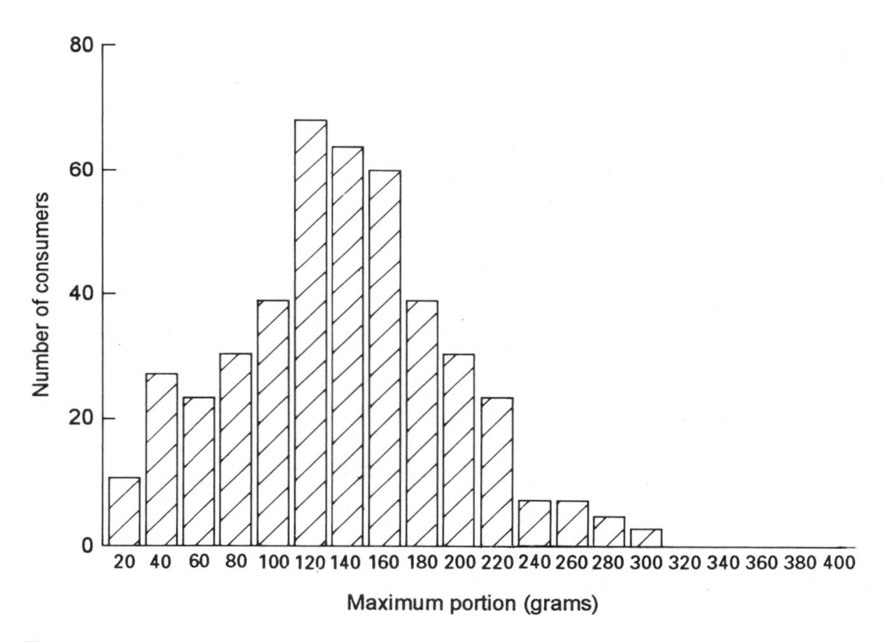

Fig. 10.2. The distribution of maximum portion sizes of oranges consumed by adults in one eating occasion.

10.3.2 Accuracy of dietary intake estimates

The accuracy of the intake estimate must be considered. For convenience, intake estimates may be classed as either actual or potential. As the name suggests, actual intakes are measured entities. For example the diet of the group of concern or a representative diet of the average consumer can be analysed. Analysing a duplicate diet of the population of concern can give a direct measure of intake and is a technique which is useful where small groups of potentially critical consumers (see section 10.3.4) have been identified. In such circumstances, an assessment of the accuracy of the results should be an integral part of the study methodology and analytical methods used. Ideally, actual intake should always be used when making comparisons with the TDI. Generally, though, insufficient data are available to allow actual intake to be determined and instead potential intakes must be estimated.

Potential intake estimates usually assume that higher concentrations are found in a wider range of foods than is realistically possible. These potential dietary intakes will invariably be overestimates of the actual intake. More realistic estimates require further information about consumption patterns, the proportion of the source food (e.g. crops) which is exposed to the chemical (e.g. pesticides) and the distribution of the residues in the source foods, and need to consider any preparation, processing or storage degradation of the contaminant residue.

10.3.3 Underlying assumptions

This third guideline states that the underlying assumptions made in order to estimate dietary intake should be acknowledged and recorded. These assumptions fall into two main areas: those made regarding the residue data, and those made for the consumption data.

10.3.3.1 Assumptions made regarding the residue data

There are at least two ways to consider the contaminant concentration in food (called occurrence assessment) and each have very different underlying assumptions:

Maximum levels allowed by law These levels are set to ensure food safety; in other words if these levels are exceeded the authorities are sure that some infringement of the law has taken place and legal action can be taken. How are these values determined? In the case of pesticides, crop trials are conducted using the maximum quantity, the greatest number of applications and with the shortest time between use and harvest, as stated on the pesticide label. The crops (in their raw state) are analysed and the highest likely residual level becomes the maximum level allowed by law (the MRL; see Chapters 2 and 8). In reality and mainly due to their cost and consumer pressure, pesticides are used at lower levels, less frequently during the growing season and not by all farmers. The levels found in crops are usually lower than the MRL and often decrease further on storage, preparation, cooking and processing. The MRL's main function is to allow for a check on good agricultural practice and they are not very helpful when estimating 'real' intakes.

Analytical levels in food It is very difficult and expensive to analyse chemicals in foods. The analytical method used in one food versus another may need to be different owing to the different compositions of foods. On analysis, some samples will have residues which are not detected by the analytical methods, and of those which are detected, the range of concentrations can be quite large. The limit of detection (LOD) can vary from parts per million (mg/kg) to parts per billion (μg/kg) or more. In other words, it is the 'cut-off' point where the chemical may exist, but the analytical test cannot measure it precisely. So what level is used in making an intake estimate? Using the maximum analysed level assumes that all the foods have the highest contamination, but if most of the samples had no detected values at all, it is not realistic to assume a person would always select the foods at the maximum. Using an average value poses another problem, do we assume the non-detected value equals the level of detection, half the detection limit or zero? Let's assume we have analysed 40 samples of food for contaminant X and have obtained the following results:

10 samples with no detectable contaminant X;
20 samples with contaminant X at the limit of detection (LOD);
10 samples with contaminant X at twice the LOD.

If we assume that the 10 samples with no detectable contaminant X really do contain no X, then the average concentration of all 40 samples would be the LOD (expressed as mg/kg or μg/kg food); if we assume they contain 0.5 times the LOD, then the average concentration for all 40 samples would be 1.125 times the LOD; and if we assume they have the LOD of contaminant X, then the average concentration would be 1.25 times the LOD. Usually, where an experimental result is less than the limit of detection, a value equal to the limit of detection is used in the estimation of intake (therefore the average would be 1.25 LOD in the example above). Sometimes the maximum is used also to give a range of results. Clearly the choice of value taken can have a significant effect on the result obtained, particularly where there are large numbers of values below the LOD.

While dietary intake estimates derived from using maximum residue levels (or other statutory limits for metals) are likely to be unrealistically high, the routine use of analytical values would be very expensive. It may be possible to start with overestimates and then use available information on storage, preparation, cooking and processing to make more realistic estimates. Previous work and experience may indicate which foods are likely to contribute the largest intake, and assist in prioritizing the analytical work. For example, let's say bread has a high concentration of a particular contaminant and levels in milk fat represent the concentrations found in all dairy products. If dairy products are consumed in relatively greater quantities by the 'at-risk' group, then it would be appropriate to analyse both bread and milk fat.

10.3.3.2 Assumptions made regarding the consumption data

There are several ways to estimate the consumption of foods, some better than others. These methods fall into two main groups, those collecting dietary information on populations and those collecting dietary information for the individual. Table 10.3 shows various methods for collecting dietary information and, where possible, the corresponding UK studies referred to in this chapter are also included.

Risk assessments are primarily concerned with the intake by individual consumers, but as these studies tend to be more expensive, methods are continuously being developed to expand the usefulness of population or household data. Most dietary studies are cross-sectional and can only reflect consumption patterns at a single point in time. If a food is consumed infrequently, like caviar, or by a small proportion of the study population, like tofu, the dietary survey is unlikely to collect sufficient information on its consumption. In order to carry out dietary intakes efficiently, several different studies are required. Table 10.4 summarizes the major sources of consumption data used in the UK at the moment. The list is expanding all the time.

There are several methodological problems which may occur when collecting consumption data. For example, there is considerable evidence to show that taking part in a dietary survey influences an individual's 'normal' eating habits. As surveys rely on

Table 10.3. Methods for collecting food consumption data from population groups and individuals in the UK

Group	Method	UK example
Individual	Food diary studies	The Dietary and Nutritional Survey of British Adults
	Duplicate diet studies	The Birmingham 1984 study
	Dietary recalls	
	Dietary histories	
Population (household)	Food diary studies	The Shipham 1979 study
	Food disappearance method	
	(i) household	The National Food Survey
	(ii) national	Per capita and production data

volunteers it is possible that the survey sample completing the study will not be a valid representation of the population as a whole.

It is likely that dietary patterns change over a person's lifetime. It would be far too expensive to follow individuals over time, in sufficient numbers to supply reliable consumption data. It is more feasible to visit households on a yearly basis. The National Food Survey (NFS) has been conducted on a yearly basis for over 50 years in the UK. National results are produced for individual months, quarters and years, enabling trends in consumption to be monitored over time. However, the NFS consumption data relates to population groups rather than individual diets. The broad pattern of the British diet during the last 50 years is remarkably consistent, considering the changes in marketing, dietary recommendations and the overall changes in lifestyle, standards of living and age structure of the population. Fig. 10.3 shows the trends in consumption of major food groups in the UK.

10.3.4 Critical groups

It is possible to have localized 'pockets' of contaminated food, so consumers living in that particular area may be at risk of higher intakes than average. Similarly the contaminant may accumulate in only a few foods—for example, mercury tends to accumulate in fish—and regular consumers of those foods may have higher intakes. It is necessary to take the geographical distribution of contaminants into account when estimating intake. Consumers living in particular locations may be a 'critical' group for a particular contaminant. As with other food chemicals, there may be groups of people, for example the old, ill or the young, who are more susceptible to the particular contaminant. Also consumers with particular dietary habits such as the very young, ethnic groups, diabetic or vegetarians, may form a 'critical' group for a particular contaminant. Special care is used to identify these possible 'critical' groups and where possible the intake assessments are focused on them.

Table 10.4. Selected dietary surveys used in the UK to estimate the intake of contaminants from foods

Study	Year	Age + size of samples	Individuals (I) or Households (H) considered	Type/design of survey	Outcomes	Comments
National Food Survey (NFS)	Yearly 1940–	All ages n = 7000/year	H	Cross-sectional nationally representative	Consumption data: 'average' of household Used to construct Total Diet Study (TDS)	Purchases estimated over a week. Food waste not corrected: data not applicable to age/sex groups. Only recently started collecting data on foods eaten outside the home.
Dietary and Nutritional Survey of British Adults	1986–87	16–64 years n = 2197	I	Cross-sectional nationally representative 7 day diary	Consumption data: Various statistics for the population and consumers. Demographic and anthropometric data	Record keeping may cause respondents to eat differently. One of the more reliable methods especially when carried out over 7 days. Literate participants necessary.
The Diets of British Schoolchildren	1983	10–11 years n = 902 boys, n = 821 girls 14–15 years n = 513 boys, n = 461 girls	I	Cross-sectional 7 day diary	Consumption data: Mean population or mean consumer data only. Demographic and anthropometric data.	Data collected on foods eaten inside and outside the home. Obtain a distribution of consumption.
Infants Survey	1986	6–12 months n = 488	I	Cross-sectional nationally representative 7 day diary	Consumption data: Various statistics for the population and consumers. Demographic and anthropometric data.	
Food Portion Sizes (Book)	1988 Amended 1990	—	—	Weighed dietary studies and manufacturers' data used to collect data on portion sizes	Portion size data	Records the average weight of the quantity of food eaten at one sitting.
Various Duplicate Diet Studies	Various	Various	I	Various studies carried out for 'at risk' groups	Consumption data for foods of interest usually grouped together.	Subjects may not correctly divide a duplicate of their diet.

As the 'critical' group may vary depending on the contaminant, we will refer to this throughout the chapter.

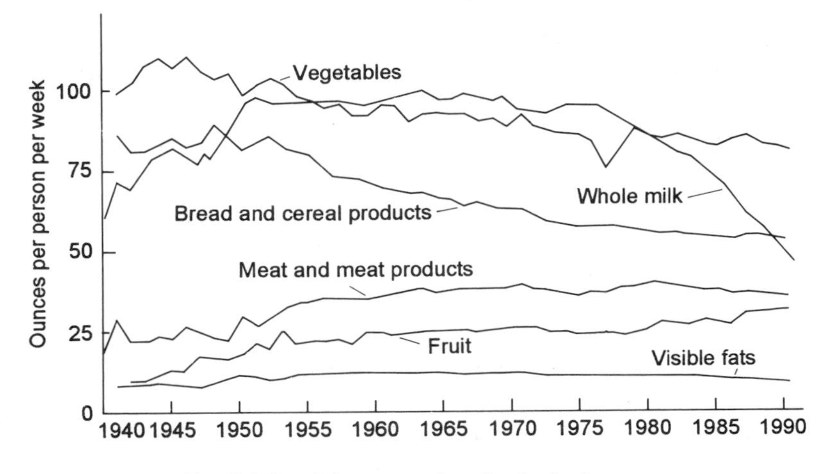

Fig. 10.3. Trends in consumption of major food groups.

10.4 HIERARCHICAL APPROACH TO ESTIMATING INTAKES

Estimating contaminant intakes from the diet is a very complicated procedure. The method chosen will depend on the toxicological properties of the contaminant (hazard evaluation) and the availability of occurrence (concentration of the contaminant) and consumption data. It is very important that before any assessment of dietary intake is made, all these pieces of information are collected, the guidelines are followed and the limitations of each of the hierarchical approaches are understood.

As we have discussed earlier, there are several ways of expressing the occurrence data and many more ways of collecting consumption data. When these pieces of information are taken together there appear to be endless options for estimating intake. Which do we use? Where do we start?

The range of food chemicals under consideration in the UK has expanded over the years. In order to respond effectively to these differing situations MAFF's Food Science Divisions have found that a variety of approaches is required. A consistent, stepwise framework has been adopted for estimating intakes of contaminants. Within this framework or hierarchical approach, some methods will be better for estimating contaminant intake (as compared to estimates for food additives for example) and some will be better suited to, say, pesticides than metals. This section outlines the hierarchical approach for estimating food chemical intakes in the UK with special reference to contaminants.

The hierarchical approach is a stepwise progression from relatively simple methods to the very sophisticated, from average estimates to 'at-risk' consumer estimates, from 'pen and paper' exercises to specifically designed studies, and from using maximum residue levels to analysed values (Table 10.5).

Table 10.5. Six basic methods for estimating dietary intakes of contaminants

Method of estimating intake	Comments
BAND 1	
Per capita	The per capita approach provides a good 'first-look' at intake averaged over the population.
Total diet study	The total diet study provides information on long-term trends in the levels of pesticides and other contaminants in the diet of the average population but not about 'critical' groups.
BAND 2	
Model diets	'Model' diets have been used to predict what intakes 'non-average' individuals might have.
Worst-case scenarios	By definition these estimates are flexible but are usually overestimates.
BAND 3	
Surveillance methods	Reliable food consumption data are required for the general population, for sub-groups within it, and for different age groups. The concentration of a chemical in the source food may be ascertained from several sources. These two types of information are combined to estimate intake.
Duplicate diets	Duplicate diet studies provide a means of looking at diets of individuals identified as potentially most at risk.

A simple first estimate can often be used to explore the likely intakes resulting from the total quantities available (per capita estimates) or from an 'average' diet (total diet study intake estimates). In some cases it may be possible to construct a range of 'scenarios' which reflect the worst possible case either in terms of food consumption, likely concentrations or both. It may also be possible to construct hypothetical or model diets. These kinds of approaches can be used as a second tier in the process. Finally, it may prove necessary to look at particular situations in detail to make a reliable estimate of dietary intake.

The collection of sufficient information to conduct an intake estimation is time consuming and expensive. The hierarchical approach offers a cost-effective stepwise and flexible framework in which to estimate the intake of food chemicals. It may not be necessary to undertake detailed and costly estimates of intake for each risk assessment. Resources can therefore be diverted to problems where more detailed analysis is required. Each approach in Table 10.5 is reviewed in turn below and, where possible, examples are given of the ways these methods have been used to estimate consumer intake of various chemical contaminants in the UK.

10.4.1 Per capita method

A per capita estimate can be used to explore the likely intakes and is particularly useful for indicating cases where the dietary intakes are high or to prioritize a long list for more rigorous investigation. Most countries have sufficient information to make per capita intake estimates. These estimates can be conducted by:

— multiplying per capita (food balance sheet) consumption by permitted levels or analysed levels for individual foods and then summing the intakes; or
— by dividing total available food chemical by the number of individuals in the population.

As mentioned previously (section 10.2.3), GEMS has constructed 'cultural' diets from the most recent FAO food balance sheets. These per capita diets have been useful to estimate possible intakes of contaminants, especially for countries with few consumption data. GEMS has also constructed a 'global' diet by using the highest average food consumption value for individual foods from each of the 'cultural' diets, adjusted to give a total of 1.5 kg/day of solid food, which allows comparison across countries.

It is not possible to estimate total available environmental contaminants, and even if it were to assume that this was evenly distributed. To produce an average figure would not be advisable. Contaminants will tend to accumulate in specific locations or foods. However, it may be possible to determine the total available pesticides and get an initial assessment of the per capita intake. However, per capita intake estimations cannot comment on variations due to different dietary habits. Per capita intake estimates are not taken on their own merit for a risk estimation.

10.4.2 Total diet study (market basket) method

Total diet studies have been carried out since the early 1960s in many countries. Initially the purpose was to estimate background exposures of the population to pesticides residues and radioactive contamination. The emphasis has shifted from pesticides to toxic metals and more recently it has included a variety of trace elements and organic contaminants. However, this method cannot be used for all contaminants. This is because the analysis of food groups may be too expensive for some contaminants and may not be feasible for others. Analytical methods may not be sufficiently reliable, the limit of detection may be too high, or the grouping of the foods (compositing) may decrease the likelihood of finding the source of contaminant.

The National Food Survey is used to construct a national average diet. Samples of each food are purchased from retail outlets throughout the country. They are combined into groups of like foods and made up to 500 g in an attempt to consider residues in minor constituents within each group. The 20 groups are prepared for eating and analysed to determine the food chemical concentration within each group. The intake from each food group is computed by multiplying the consumption by the concentration. This is added to give the total average intake.

Total diet studies have been carried out regularly in Britain since 1966 and used to estimate the average intakes of pesticides or other contaminants from the diet as a whole. Since the 1960s the study has been modified, firstly in 1975 and more extensively in 1981. It now includes 115 foods in 20 food groups compared to 68 foods in nine groups in the previous design, and reflects more recent food consumption patterns. Table 10.6 shows the changes made to the UK total diet study over a number of years.

Table 10.6. Total diet (market basket) studies in the UK

Year	Group	No. of food items (groups)	Reference†
1966–1975	Average person	78(11)	Harris *et al.* (1969)
1976–1981	Within a household	68(9)	Buss and Lindsay (1978)
1981–present		115(20)	Peattie *et al.* (1983)

† See Further Reading at the end of this chapter.

Total diet studies work in the following way. Let us assume a pesticide called CHEM-X is approved for use on apples, oranges and potatoes. We want to estimate the average intake of CHEM-X for the general population, so we conduct a total diet study concentrating on the two food groups of interest. The fresh fruit group includes oranges, other citrus fruit, apples, pears, other stone fruit, bananas and other fresh fruit. The potato group includes fresh potato and potato products. We assume the mean analysed levels of CHEM-X are 0.1 mg/kg in fresh fruit and 0.7 mg/kg in potatoes. Table 10.7 shows the intake of CHEM-X from these two foods groups. We will use CHEM-X again to illustrate other methods of estimating dietary intakes, but as the residue values are made up, any comparison across the methods will be superficial.

Table 10.7. Dietary intake of CHEM-X (mg/person per day)

Food†	Analysed residue level (mg/kg)	Estimated weight of food eaten (kg/person per day)	Intake (mg/person per day)
Fresh fruit	0.1	0.060	0.006
Potato	0.7	0.159	0.1113
All foods		0.219	0.1173‡

† Fresh fruit includes oranges, other citrus fruit, apples, pears, other stone fruit, bananas and other fresh fruit. The potato group includes fresh potato and potato products.
‡ Assuming the average bodyweight is 60 kg, this intake is approximately 2 μg/kg bodyweight per day.

There are several advantages in conducting a total diet study:

— it provides information on long-term trends in the levels of pesticides and other contaminants in the diet;

— it highlights the food groups which contribute the most to the overall average intake— so it can be used to prioritize further surveillance work;
— it is a cost-effective way of determining contaminant intake by the average person;
— it provides a baseline reference for determining the impact of environmental accidents or regulatory control on the food supply;
— as the total diet study represents the intake resulting from an average diet it has been used to express daily 'background' intake when considering local contamination.

There are several limitations of the total diet study:

— it does not provide information about individual consumption patterns, or the distribution of intakes;
— the food groups are widely defined so that the major source foods cannot be determined;
— it cannot be used to express intakes by 'critical' groups.

10.4.3 Model diet method

'Model' diets have been used to predict what intakes 'non-average' individuals might have, to determine whether all of the population is protected by nationally or internationally agreed safety standards. This method serves a useful purpose in the interim period between the time the contaminant is identified as a potential problem until sufficient information is available to make better assessments.

In the simplest estimates the intake of the contaminant from one serving of a source food can be expressed as a percentage of the TDI. This method is particularly useful when few consumption data exist or when the major source of the contaminant is one food. 'Modelled' intakes are not taken on their own merit for a risk estimation.

A 'model' diet is currently being used to establish permissible MRLs for veterinary drugs residues while ensuring that the ADIs are not exceeded by consumers. A hypothetical adult diet was proposed by the Joint FAO/WHO Expert Committee on Food Additives and confirmed in 1990. In order to protect all segments of the population, exaggerated consumption values were used to construct a 'model' diet. The diet comprises:

300 g lean muscle
100 g liver
50 g kidney
50 g animal fat
1.5 litres milk or milk products
100 g eggs
and 20 g honey.

This has been compared to the consumption data collected from the UK Dietary and Nutritional Survey of British Adults on the basis of the weight and calorie content. Preliminary work suggests that the hypothetical diet contains at least twice the amount of animal products eaten by the 'top-centile' adult consumer and supplies more calories (from these animal products) than expected for the whole adult diet. In other words the

MRLs estimated using this 'exaggerated' diet are probably lower than they need to be to ensure safety to the consumer.

The disadvantage of the above 'model' diet is that it was designed to consider the possible sources of veterinary drugs so it is not very useful when considering food chemicals which may, for example, be used as a veterinary drug as well as a pesticide.

10.4.4 'Scenario' method

In some cases it may be possible to construct a range of 'scenarios' which reflect the worst possible case either in terms of food consumption, likely concentrations or both. This is a slightly more refined method than 'modelling' as it assumes more information is available. This approach attempts to extend the usefulness of the data and predict trends or intakes of 'at-risk' groups.

There is scope to incorporate 'adjustment' factors in order to make the estimates more realistic. These 'adjustment' factors may include, for example differing levels of migration if considering the potential intake of a chemical from a packaging material, varying preparation habits if considering pesticide intake, estimating the proportion of liver which is derived from a specific species if considering veterinary drug residues and so on.

There are two main applications of this method:

— *Where sufficient information is available but from very different sources.* 'Scenario' estimates allow the 'mixing' and 'matching' of different information databases. Consumption data from one dietary survey are very unlikely to represent all 'at-risk' groups and may not include all the foods required. Several studies may be used to construct an overall picture of consumption. Similarly residue data may be generated in different years, by different laboratories, for different reasons. The available information is used, where possible, to represent the likely concentrations. Clearly judgement and experience are needed and these evaluations are subject to error. The 'scenarios' are usually based on 'worst-case' assumptions so are more likely to be overestimates than underestimates.

— *Where only a very expensive, lengthy and very specialized study would estimate the intake.* The best example of this is the case of migration of contaminants into food. If a food chemical is slowly leaching from a vessel into a food, the rate of leaching generally decreases over time and re-use of the vessel. Some people consume greater or lesser quantities of the food than the average. How do you estimate the consumer intake? This is a rather interesting problem. Assuming the food is stored in the vessel until it is consumed, an infrequent consumer may have food with progressively higher concentrations of the chemical. However, their consumption will be relatively low if averaged over time. On the other hand, a frequent consumer may consume more but, as the food spends less time in contact with the vessel, at a lower concentration. It may be that the average consumer is at risk of the highest intake averaged over time. If there is a concern that this intake exceeds the safety standard an assessment is required quickly and one of the best ways is to use the 'scenario' approach. So for this particular example, the frequency of consumption for the 'light', 'moderate' and 'heavy' consumer is considered with the available leaching data. Intake is estimated on a weekly basis for each consuming group. This *theoretical* exposure is continued for a year or two (depending on the expected use of the vessel and availability of leaching data). Average intake over a period of time can then be evaluated using 'worst-case' assumptions.

10.4.5 Surveillance methods

It may prove necessary to look at particular food chemicals in detail. Reliable food consumption data are required for the population of interest and the data are then combined with residue data. If the contaminant is used on a limited range of foods then it may be possible to identify a particular sub-population who are consumers of these foods and to conduct a targeted survey of their dietary habits.

A number of years ago, the soil concentrations of zinc, lead, copper and cadmium around the village of Shipham in Somerset, UK, were found to be high. Three dietary surveys (two diary studies and one duplicate diet study) and a crop sampling survey were carried out in 1979 to determine what the dietary intakes of these metals were for the people living in the area. Average weekly consumptions of fresh fruit and vegetables per household over a period of four weeks coinciding with the growing seasons (May and September), were determined from the diary studies. Total intake was then estimated by simply adding intake from the rest of the diet (as determined by the British total diet study). The duplicate diet study (in September, 1979) was used to estimate the total intakes of the metals by individuals. Table 10.8 shows the results of these surveys with the corresponding tolerable intakes for the metals.

The main findings from this study were that the intakes of lead, zinc and copper by residents of Shipham were very similar to the average intakes by the UK population. There was evidence that the cadmium intake was higher than average and a few residents exceeded the tolerable intakes. However, given that the studies were conducted at times of year when consumption of home-grown fruits and vegetables was high, intake was very unlikely to be maintained at other times of the year. However, as a precaution, residents were advised to avoid growing and eating locally grown crops which contained raised concentrations of cadmium (potato, cabbage, spinach, lettuce and kale). The concentrations of lead and cadmium were high in soil so it was advised that all vegetables were washed thoroughly before eating and pica (soil eating by children) was strongly discouraged. Smoking also contributes to cadmium intake (20 cigarettes a day can increase the average intake by 50%), so residents were advised to switch to pipes or cigars, or quit smoking.

Food chemicals may have several functions and some contaminants may be found in a wide range of foods. It would not be feasible in such circumstances to conduct a specific diary study for each food additive or contaminant. In these circumstances national dietary surveys of individuals are more appropriate.

The Dietary and Nutritional Survey of British Adults (British Adult Study) has been used extensively to estimate intake data for the 'top-centile' consumer. The study was carried out between 1986 and 1987 on behalf of MAFF and the UK Department of Health. A nationally representative sample ($n = 2197$) of adults aged 16–64+ completed a seven day, weighed, diary record of foods consumed inside and outside the home. The study employed over 5000 food codes. The British Adult Study is computerized and with a recipe database can be interrogated to estimate consumption of single or groups of

Table 10.8. Dietary intake of metals at Shipham (mg/person per week, mean value and range)†

Metal intake	Cadmium	Lead	Zinc	Copper
Dietary estimate‡ based on records kept by 75 families during periods of four weeks in May and September 1979	0.23 (0.11–0.50)	0.71 (0.6–1.1)	72.5 (66.1–111)	10.4 (9.0–13.6)
Duplicate diet estimate based on analysis of duplicates of a single week's diet, provided by 65 individuals during September 1979	0.18 (0.007–1.06)	0.42 (0.14–1.2)	61.0 (38–100)	7.3 (3.8–17.8)
National average estimated based on 1978 data and reflecting average food consumption in the UK	0.10(5) (0.05–0.17)	0.602 (0.2–1.2)	70.0 (56–86)	11.3 (7.8–14.6)
Provisional tolerable weekly (PTWI) or tolerable daily intake (TDI)§	0.42	3.0	420	210

† The estimates include the contribution from all dietary components, including milk and beverages.
‡ These results are overestimates because of assumptions made in the calculations.
§ Assuming an average adult bodyweight of 60 kg.

foods. If the concentrations of the contaminant are entered within the computer program, the potential intake can be calculated.

As an example, let's assume that CHEM-X is a pesticide registered for use only in oranges, potatoes and apples and that the MRLs are 2 mg/kg for oranges, 0.2 mg/kg for potato and 0.5 mg/kg for apples. The mechanics for estimating the intake from the British Adult study are as follows. The daily orange consumption (in kg) for each person is multiplied by 2, the daily potato consumption for each person is multiplied by 0.2 and the daily consumption of apples for each person is multiplied by 0.5. For each person on the database, the total intake is the sum of intakes from the three commodities. If a given individual did not eat oranges, potatoes or apples during the week of the study, their intake would be zero and not included in further analysis. The computer program then constructs a distribution of the individual intakes (expressed by bodyweight) and from this distribution the mean, 'top-centile' and maximum intakes can be extracted (see Table 10.9). For this example the estimated intake of CHEM-X of the mean, 'top-centile', and maximum consumer are 0.8, 2.8 and 6.0 μg/kg bodyweight per day respectively.

Table 10.9. Dietary intake of CHEM-X

		Intake of CHEM-X (μg/kg bodyweight per day)‡		
Food†	Maximum residue level (mg/kg)	Intake by mean consumer	Intake by 97.5%ile consumer	Intake by maximum consumer
Oranges	2.0	0.7	3.1	5.1
Potatoes	0.2	0.4	0.9	1.6
Apples	0.5	0.3	1.0	3.2
All foods		0.8	2.8	6.0

† A recipe database has been used to include consumption of the food commodity within dishes, for example orange in fruit salads, apples in apple pie and potatoes in shepherds pie.
‡ Different groups of consumers are considered for each food, so the intake cannot be added to estimate the intake from all the foods.

10.4.6 Duplicate diet method

Duplicate diet studies (DDS) are a variation on the total diet study approach. The DDS provide a means of looking at diets of individuals identified as potentially most 'at-risk', either because they are thought to be exposed to a higher level of contaminant than the average consumer or because they are known to consume more of the source foods.

Individuals most 'at-risk' from high dietary intakes of food contaminants are requested to supply a duplicate of their meals to be analysed. Analyses of diets are valuable in determining the dietary intake of contaminants from foods prepared as they would be eaten. However, the participants may not correctly divide the meals between themselves and the 'duplicate plate', and it is more difficult to distinguish the major sources of contamination.

For many years there has been concern about the intake of lead by infants, so much so that a separate PTWI (25 μg/kg bodyweight per week) has been set for this age group, which was half that of adults (50 μg/kg bodyweight per week). A seven-day duplicate diet study was conducted with 97 pre-school age (24–27 months old) children from the Birmingham area in 1984. Their mothers were asked to collect and weigh duplicate samples of all food and drink consumed during a week. These samples were placed into two containers (one for food, one for the liquids) and analysed for lead and other metals. The average amount of food consumed per week was 2.6 kg/week (0.4–5.4 kg/week), and for beverages 4.12 kg/week (1.2–8.4 kg/week). Table 10.10 shows the estimated dietary intakes for selected heavy metals studied in Birmingham by the duplicate diet study. In Birmingham, 9% of the children's intakes exceeded the PTWI for lead, but none of these children exceeded the 25 μg/dl advisory action level in blood. The PTWIs for cadmium, copper, zinc and tin were not exceeded by any of the participants.

Table 10.10. Dietary intake of metals in Birmingham UK (mg/person per week, mean value and range)

Metal intake	Cadmium	Lead	Zinc	Copper
Duplicate diet estimate based on analysis of duplicates of a single week's diet, provided by 97 children. 1984.	0.035 (0.01–0.1)	0.185 (0.06–1.1)	22.9 (6.8–40.8)	3.15 (1.4–18.5)
Provisional tolerable weekly (PTWI) or tolerable daily intake (TDI)†	0.084	0.3	84	42

† Assuming the average bodyweight of the young children is 12 kg.

10.5 ESTIMATED CONSUMPTION OF MILK OVER A LIFETIME

Tolerable intakes are expressed as averages over a lifetime, but consumption data are usually collected on a specific age group over a relatively short period of time (usually four or seven days).

Infants consume milk as a major part of their diet. If a contaminant is found in whole milk the intake will be high in this age group, but the average value over a lifetime will be lower as adults consume substantially less whole milk (especially if you consider their greater bodyweight). So what is the consumption of milk averaged over a lifetime?

Fig. 10.4 illustrates the life-time consumption of whole milk by the 'top-centile' consumer. The 'top-centile' consumption of whole milk by adults is 550–620 g/day (average 576 g/day), but closer to 1 litre (approximately 1000 g) for infants. The average 'top-centile' consumption over a lifetime has been estimated as 614 g/day. This is approximately a pint a day, *every day of your life*. The consumption data used to generate Fig. 10.4 were extracted from different dietary surveys. Several assumptions were used. For example, we assumed that people stopped consuming milk at age 70 and the data collected for the two-to-five-year olds were too unreliable to use, so it is subject to error and criticism.

A rolling programme of diary-based surveys is being commissioned jointly by MAFF and the UK Department of Health, and the next to be completed study is on toddlers aged $1\frac{1}{2}-4\frac{1}{2}$ years. Following this the elderly and school children will be surveyed. However, it will be several years before we have sufficient information to test how close the estimated 'lifetime' milk consumption value is to the survey value. The 'real' value can only be estimated by following individuals throughout their lifetimes.

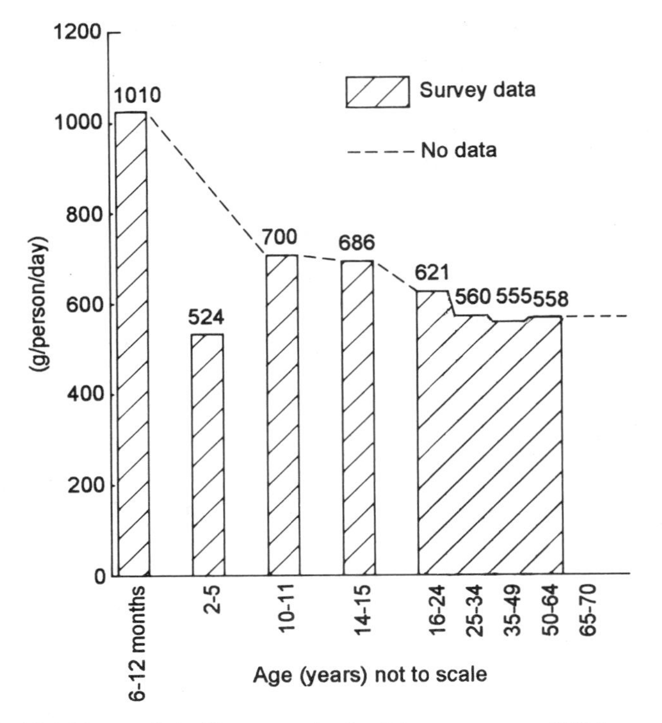

Fig. 10.4. 'Top-centile' milk consumption by different age groups (UK figures).

10.6 INTERNATIONAL GUIDELINES FOR ESTIMATING THE INTAKE OF CONTAMINANTS

International guidelines for pesticides and other chemical contaminants have adopted a progressive or 'hierarchical' approach, and have provided a useful framework in which to compare dietary intake estimates across the European Community.

10.6.1 Pesticides

The Codex Committee on Pesticide Residues has developed guidelines for estimating pesticide intake using *per capita* consumption—the 'cultural' diets and the 'global' diet have been mentioned previously (sections 10.2.3 and 10.4.1). These guidelines suggest the first estimate should use the 'global' or national diet and assume that all source foods contain the MRL to calculate the theoretical maximum daily intake, TMDI, and then use a 'cultural' or national diet with any possible processing losses to derive an estimated maximum daily intake, EMDI. Finally intakes should be estimated using national consumption data and surveillance data or crop use to give the estimated daily intake, EDI. In other words, these three steps for estimating pesticide intake start by using a global diet, assume that all foods contain the MRL, and end by estimating intakes using national consumption data and surveillance data. Table 10.11 shows the TMDI, EMDI and EDI intakes of CHEM-X, assuming the same MRLs as before (p. 175). For simplicity, we also assume that there is 50% processing loss in each food and the pesticide CHEM-X is used on only 50% of the crops. The guidelines allow different reduction factors for the different crops if necessary.

Table 10.11. Dietary intake of CHEM-X using the global and European diets

Food	MRL (mg/kg)	Intake of CHEM-X (μg/person per day)	
		Global diet	European diet
Oranges	2	40.0	98.0
Potatoes	0.2	17.2	41.2
Apples	0.5	11.5	27.5

Dietary intake of CHEM-X (μg/kg bodyweight per day) using WHO guidelines:

$$\text{TMDI} = \frac{(17.2 + 40.0 + 11.5)}{60}$$
$$= 1.1\,\mu\text{g/kg bodyweight per day}$$

$$\text{EMDI} = \frac{(41.2 + 98.0 + 27.5) \times 0.5}{60}$$
$$= 1.4\,\mu\text{g/kg bodyweight per day}$$

$$\text{EDI} = \text{EMDI} \times 0.5 = 0.7\,\mu\text{g/kg bodyweight per day}$$

10.6.2 Other contaminants

The Joint FAO/WHO Expert Committee on Food Additives has evaluated a number of food contaminants and established provisional tolerable weekly intakes (PTWIs). The Joint UNEP/FAO/WHO Food Contamination Monitoring Programme or GEMS/Food provides information used in these safety considerations. Thirty-nine countries participate in GEMS and the information collected on contaminant intake has been derived from total diet 'market basket' studies. The composition of the diet and preparation for analysis vary from country to country but allow the monitoring of trends within each country. Contaminants studied include certain organochlorine and organophosphorus pesticides, PCBs, cadmium, mercury and lead. The summary from a recent report states that:

> 'In most cases, dietary intakes of organochlorine and organophosphorus pesticides are well below the Acceptable Daily Intake (ADI) of the respective pesticide. Of some 21 countries providing information on the average dietary intake of cadmium, only in one case is the Provisional Tolerable Weekly Intake (PTWI) exceeded. Several countries identified cereals and cereal products and root and tuber

vegetables as the main contributors to the dietary intake of cadmium. For mercury, all reported intakes are below the PTWI of methylmercury. The contribution of fish to the total intake of mercury varied from 20 to 85 per cent, depending on the country. Therefore, the general assumption that fish is the main contributor to the total dietary intake of mercury may at times, not be justified. Average dietary intakes of lead exceeding or approaching the PTWI are reported for adults and infants and children in some countries. Foodstuffs which contribute most to the intake of lead vary from country to country, and have been identified as being drinking water, beverages, cereals, vegetables and fruit.'

10.7 CONCLUSIONS

In summary, the estimation of intakes of food chemicals is a complex activity and no single approach is suited to all circumstances. A variety of approaches are available for estimating intakes of food chemicals and care must be taken to ensure that an appropriate method is used. Six basic approaches have been discussed with examples. Each case assessed requires a specially tailored approach according to the data available and the use to which the result will be put. The approaches employed are designed to set consumer safety as the highest priority.

Various dietary survey techniques employed by MAFF are illustrated. The cross-sectional, weighed food diary approach has proved exceptionally useful in studies where intake by 'non-average' consumers is required. Separate dietary surveys or duplicate diet studies are more suited to consider intakes of 'critical' or 'at-risk' groups.

The area of risk assessment is developing very quickly. Further research is required not only to investigate the many underlying assumptions and uncertainties but also to develop better methods for estimating the intake of contaminants in 'at-risk' groups.

FURTHER READING

MAFF's role in food safety

Atkins, D. P. and Smith, S. M. *British Food J.*, **91**, 15–23. (1989)

Knowles, M. E., Bell, J. R., Norman, J. A. and Watson, D. H. *Food Additives and Contaminants*, **8**, 551–564. (1991)

Sources of consumption data

Gregory, J., Foster, K., Tyler, H. and Wiseman, M. *The dietary and nutritional survey of British adults*. HMSO, London. (1990)

Household food consumption and expenditure. Annual Report of the National Food Survey Committee. HMSO, London. (1989)

Ministry of Agriculture, Fisheries and Food. *Fifty years of the National Food Survey 1940–1990*. HMSO, London. (1991)

Report on Health and Social Subjects No. **36**: *The diets of British schoolchildren*. HMSO, London. (1989)

Ministry of Agriculture, Fisheries and Food. *Food and nutrient intakes of British infants aged 6–12 months*. HMSO, London. (1992)

Intake estimations

Buss, D. H. and Lindsay, D. G. *Fd. Cosmet. Toxicol.* **16**, 597–600. (1978)

Harris, J. M., Jones, C. M. and Tatton, J. O'G. *J. Sci. Fd. Agric.* **20**, 242–245. (1969)

Peattie, M. E., Buss, D. H., Lindsay, D. G. and Smart, G. A. *Fd. Chem. Toxic.* **21**, 503–507. (1983)

Smart, G. A., Sherlock, J. C. and Norman, J. A. *Food Additives and Contaminants*, **5**, 85–93. (1987)

Sherlock, J. C., Smart, G. A., Walters, B., Evans, W. H., McWeeny, D. J. and Cassidy, W. *The Science of the Total Environment*, **29**, 121–142. (1983)

International guidelines on estimating dietary intakes

Joint FAO/WHO Expert Committee on Food Additives *Evaluation of certain veterinary drug residues in food.* WHO Technical Report Series, Report **34**, no. 788. (1989)

WHO Guidelines for Predicting Dietary the Intakes of Pesticide Residues. Report prepared by the Joint UNEP/FAO/WHO Food Contamination Monitoring Programme in collaboration with the Codex Committee on Pesticide Residues. World Health Organization, Geneva. (1989)

WHO Guidelines for the Study of Dietary Intakes of Chemical Contaminants. Report prepared by the Joint UNEP/FAO/WHO. World Health Organization Offset Publication No. **87**, Geneva. (1985)

11

Future scientific work on chemical contaminants in food

D. H. Watson, Ministry of Agriculture, Fisheries and Food, R242, Ergon House, c/o Nobel House, 17 Smith Square, London SW1P 3JR, UK.

11.1 INTRODUCTION

Although much is known about the various groups of chemical contaminants in food, there are several important areas in which fundamental work remains to be done. There are several important questions to be answered. These include the following:

— Are there any unrecognized chemical contaminants in the diet which may pose a risk to the consumer?
— How representative, of the food supply under study, are the food samples we use in surveillance for chemical contaminants in food?
— How sensitive should methods of analysis be?
— What is the significance, in risk management, of (a) mixtures of chemical contaminants in food, and (b) chemical contaminants binding to macromolecules in food?

Although none of these questions is new, they remain largely unexplored. The questions are important, particularly in organizing and assessing the growing body of work on chemical contaminants in food. This scientific area is now moving from the cataloguing of chemicals, which has taken over two decades, towards a systematic approach to assessing the risks from their presence in food. For this assessment and the control of chemical contamination of food to be fully effective, the types of questions posed above need to be answered. As the questions are not new, and yet they remain largely unanswered, new approaches are probably needed to make progress. This chapter reviews the above questions and suggests some new approaches that might be used to develop answers to them.

11.2 DISCOVERY OF CHEMICAL CONTAMINANTS IN FOOD

So far several hundred chemicals contaminating food have been discovered. Many of these, notably some mycotoxins and industrial chemicals, have been found in relatively few samples of food, perhaps because not many samples have been analysed. It is not clear if these substances are found widely in food. Clearly it is important to answer this point, but because there are so many chemicals involved and there are many different foods, this would need very effective and extensive use of food surveillance resources. This might be achieved by the analysis of combined samples of like foods to reduce the amount of analytical work involved. But since combining samples usually leads to the dilution of contaminants, analysing composite samples would only save resources where there are high levels of contamination and/or an even distribution of the chemical contaminants in the tested sector of the food supply. The latter assumption is discussed below (section 11.3). As regards the degree of contamination, if it is not at a high level it is unlikely to be of major interest unless the contaminant is particularly toxic. Indeed if considerable or extensive contamination is found, this is likely to lead to some toxicological work on the chemical involved, and this in turn may lead to further analysis for it in food.

A pragmatic approach is required if the currently extensive but tentative list of chemicals that might contaminate food is to be shortened. The alternative, more established approach, of experimental work that concentrates on the detail of the chemical and toxicological properties of contaminants, is likely to add to the list of putative contaminants without addressing the main issue of whether the chemicals involved contaminate food frequently, rarely or at all. It is not proposed that such detailed work is reduced. Instead it is suggested that there is further work to develop broad screens for chemical contamination of the food supply, with particular emphasis on searching for chemicals that previous work has indicated might be present but on which there has been little or no experimental work. There has been some work already on broad screening methods. Attempts to use biological systems, such as mutagenicity tests, have had relatively little impact because they do not identify the biologically active chemicals. Chemical methods probably offer more, provided they can be developed to detect a wide range of chemical contaminants and particularly to analyse a wide variety of foods.

There is also a need to search for any new *groups* of chemical contaminants in food. Since the 1960s and 1970s when the currently accepted groups of food chemical contaminants were established—for example residues of drugs and pesticides, chemicals migrating from plastic, heavy metals and mycotoxins—relatively few new groups of chemical contaminants have been identified. Very general classifications, particularly that of 'environmental chemicals', may mask several groups of chemicals which are of major importance in assessing risk from food chemical contaminants. It is debatable whether the broad screen approach suggested above would be of help here since new groups of chemical contaminants in the food chain may differ quite radically from known ones. A new approach is needed to identify such groups of chemicals. One potentially fruitful means could be to look at the major environmental input of stable organic chemicals not so far detected in food, and to test marine biota, soil and herbage for contamination with candidate chemicals. A straightforward method is needed to select those chemicals that are most likely to persist in the food chain. (A possible way of proceeding is described in Chapter 3, section 3.1.) Coupled with information about which chemicals are used and might be released in a given locale, environmental contamination by a selected list of chemicals could be used to steer work to identify possible contamination of the food chain by those groups of chemicals that are released into the environment.

Given the growing trend for 'natural' foodstuffs, it is also worthwhile considering the major gaps in our knowledge about the occurrence of naturally occurring toxicants in food. Too little is known about the extent to which many toxicants occur in food. The other major gap in our knowledge is about toxicants produced by animals. Work on natural toxicants has concentrated on micro-organisms, notably fungi, as these are associated with crop disease and food spoilage. However, work on higher plants has shown that natural toxicants are also produced by some food-producing plants. The same might be true of some food-producing animals.

11.3 REPRESENTATIVENESS OF FOOD SAMPLES

Our knowledge of chemical contamination in food is largely based on information about the incidences and levels of chemical contamination in samples of food and food raw materials. Although analytical methods have become increasingly sensitive and exact, sampling plans used in obtaining foodstuffs for analysis have not been developed to a similar degree of sophistication.

Whilst highly sophisticated sampling plans may not always be necessary—for example where most samples of a homogeneous foodstuff are evenly contaminated with a chemical—there is little point in analysing samples that are not representative of the general supply. The following questions are amongst those one needs to consider in developing representative sampling plans.

— What is the likely distribution of the chemical contaminant in the supply of the given foodstuff?
— What is the best place in the food supply chain to obtain representative samples?
— How many samples are required?

These questions are considered below (sections 11.3.1 to 11.3.3).

11.3.1 Contaminant distribution

It cannot be assumed that chemical contaminants will be normally distributed in supplies of food and related materials—by 'normal' I refer to a bell shaped, Gaussian or normal distribution. Fig. 11.1 shows skewed and probably discontinuous distributions of chemical contaminants in several extensive surveys in this country. Similar distributions were found in 13 out of 15 surveys examined in this way. The surveillance covered a wide range of chemicals—including organophosphorus and organochlorine pesticides, polychlorinated biphenyls (PCBs), sulphadimidine, radiocaesium, di-2-ethylhexyladipate, aflatoxins, nitrate and lead. The range of commodities was also quite wide: bran, pork products, lamb, potatoes, kidney, chicken, peanut butter and lettuce. (In one survey of PCBs human fat was monitored.) In only two cases—radiocaesium in a flock of sheep, and PCBs in human fat—the results fitted a normal distribution (Fig. 11.2). Both of these

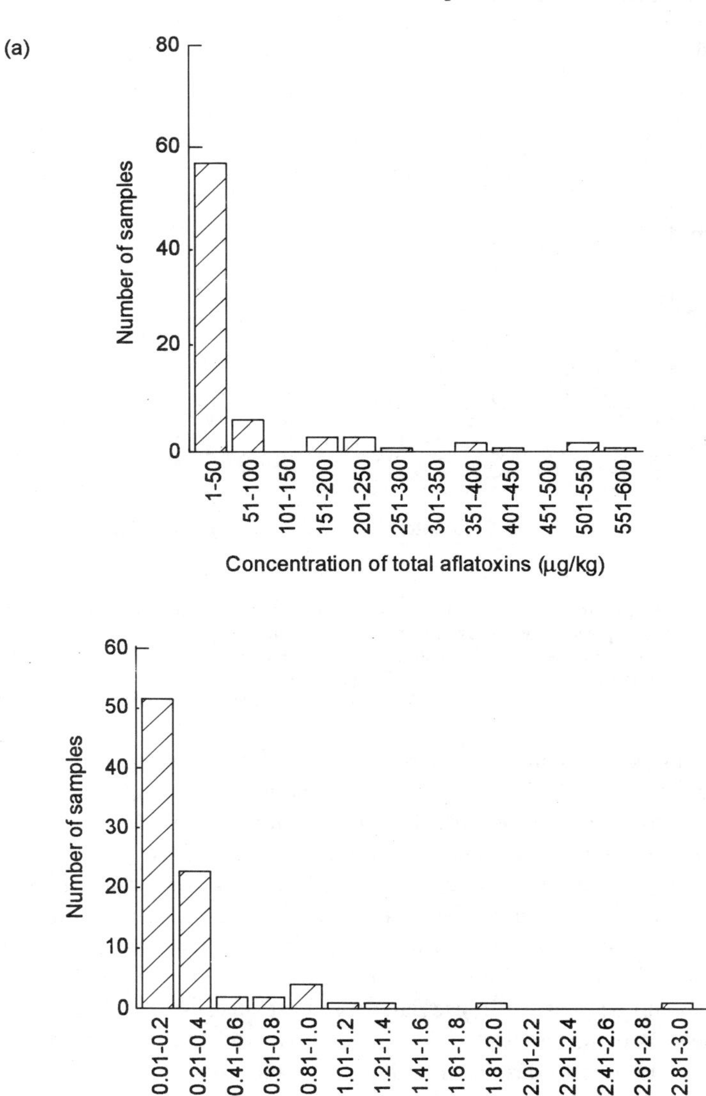

Fig. 11.1. Skewed distributions of some chemical contaminants in samples of food. In each case the distribution of positive results is shown. Details of how the surveys were carried out are given in the source references. (a) Total aflatoxins in peanut butter samples. Details of this survey are given in Table 3 of Food Surveillance Paper No. 18, published by HMSO, London, 1987 (samples were of crunchy and smooth peanut butters from health food producers). The results in the bar chart are for 78 samples in which one or more aflatoxins (B_1, B_2, G_1 and G_2) were detected. (b) Pirimiphos-methyl in bran. Details of this survey are given in Table 15 of Food Surveillance Paper No. 25, published by HMSO, 1989. The results of the histogram are for 87 samples in which pirimiphos-methyl was detected.

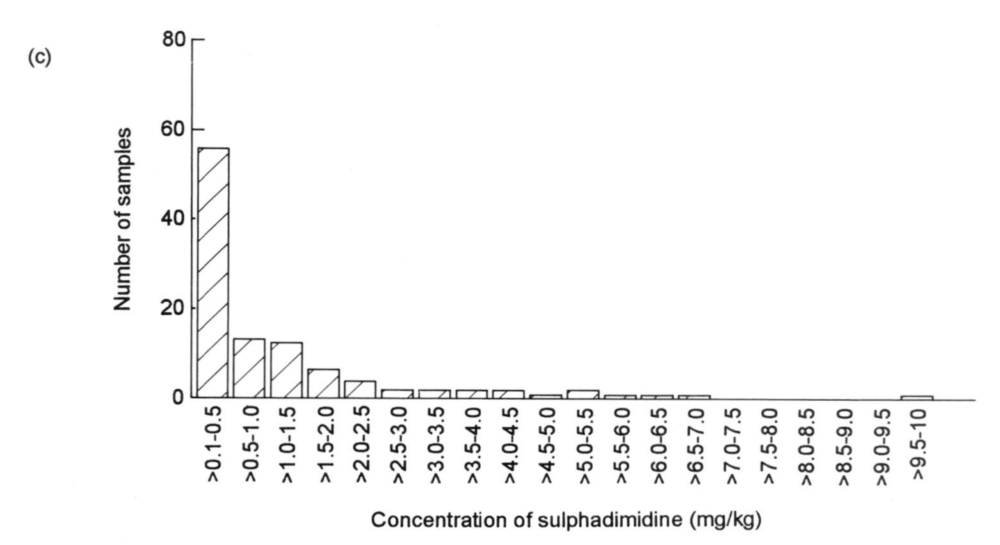

Fig. 11.1. Skewed distributions of some chemical contaminants in samples of food. In each case the distribution of positive results is shown. Details of how the surveys were carried out are given in the source references. (c) Sulphadimidine in pig kidney. Details of this survey are given in Table 5 of Food Surveillance Paper No. 33, published by HMSO, 1992. The results in the graph are for 110 samples with residues above 0.1 mg/kg.

normal distributions were understandable. The results for radiocaesium were for one lock of sheep whilst the PCBs data were for residues following exposure over a long period of time. Whilst the skewed distribution of aflatoxins (Fig. 11.1(a)) could also have been anticipated—this is well established for these toxins—the non-Gaussian distributions for the other contaminants were unexpected. Should skewed distributions apply in the majority of surveys carried out on chemical contaminants in food, this would clearly need to be considered in designing sampling plans. In this circumstance it would be wrong to assume that one is dealing with a homogeneous distribution of chemical contaminants. It should be possible to use data from past surveys to build up a fuller picture of the distributions of different chemical contaminants in the many sectors of the food supply. This would help to design sampling plans with more confidence. It should also aid the reporting of results since means and/or medians would need to be used where appropriate, and preferably the actual distribution of results in a survey should be given as well.

11.3.2 Choice of sampling point

In theory the best site to obtain representative samples is the point (or points) where the various parts of the supply chain converge. For example, sampling of meat and offal is more likely to be representative of farms in a district if samples are obtained at the local slaughterhouse as, in theory, the flow of animals represents the overall supply from the locale. In practice it is sometimes not known if this is the case as farmers may transport animals to a variety of slaughterhouses, some quite far from their farms. It is certainly a

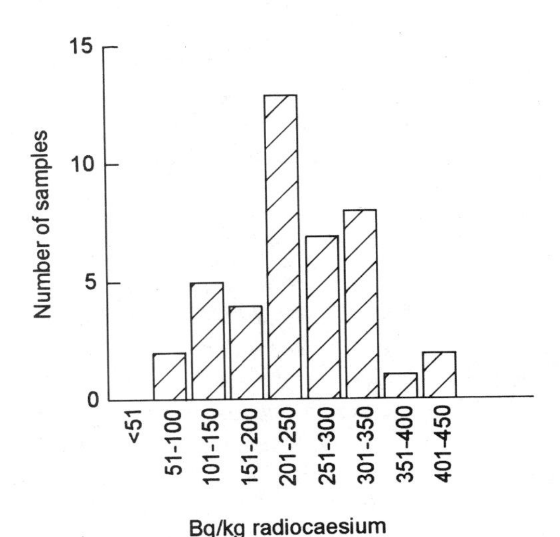

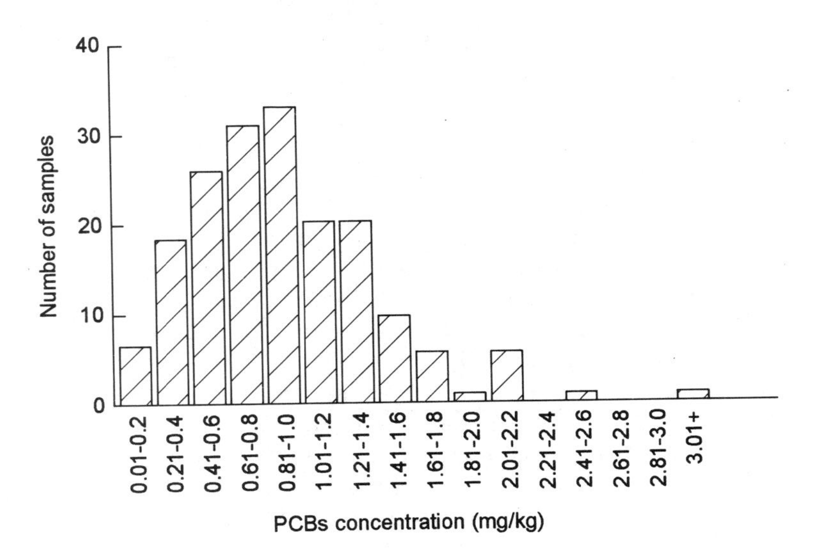

Fig. 11.2. Normal distributions of some chemical contaminants in samples of food and other materials. (a) Radiocaesium activity in sheep from one farm. This is part of extensive monitoring of radioactivity in food in the United Kingdom (see Food Surveillance Paper No. 28, published by HMSO, London, 1990). The results in the histogram are for 42 samplings. (b) PCBs in samples of human fat. Details of this survey are given in Table 7 (1982 to 83 results) of Food Surveillance Paper No. 16, published by HMSO, 1986. The results in the bar chart are for 186 samples in which PCBs were detected.

more effective use of resources to obtain samples where most of the supply chain converges but whether this leads to truly representative samples is not known. Indeed little is known about the effects of sampling at different points in the food chain, although there is some evidence that processing can effect the distribution of contaminants in raw materials. Further work on this would be helpful, particularly on the distribution of a given chemical contaminant as it passes through the food chain.

11.3.3 Numbers of samples

Given the shortage of information about how to obtain representative samples, not surprisingly little is known about how many samples are needed to provide a representative picture of the levels and incidences of chemical contamination of food. This is despite efforts to establish the minimum number of samples required to detect specific incidences of contamination (see for example Chapter 2, section 2.4.3), usually on the basis of statistically-based models derived from epidemiological studies of disease. The shortage of information about the distribution of chemical contaminants in the food supply (section 11.3.1) is a major problem in deciding whether epidemiologically-derived models are appropriate for the design of sampling schemes for food chemical surveillance. Considerable work would be needed to test this. At present one must rely on a pragmatic approach (Table 9.3). Analytical resources usually constrain a survey to 100 or less samples per analyte (or analyte group if several chemical contaminants are analysed per sample). Experience shows that usually 50 to 100 samples provide a reasonably extensive picture of the incidence and levels of a contaminant in a commodity, but this is only a rough guide and is probably only appropriate if more than a few per cent of samples are chemically contaminated.

11.4 SENSITIVITY OF ANALYTICAL METHODS

There has been considerable progress, over the last 20 years, in detecting the usually very low levels of chemical contaminants in food. Detection at levels of parts per million is now being superseded as normal practice by analytical sensitivities of one part per billion (10^9) or less. This inevitably raises the question of how sensitive analytical methods must be. The answer depends on circumstances. For example it is certainly not cost effective for regulatory authorities or industry to have a highly sophisticated method of analysis if standards set by legislation can be achieved by using a simpler, less sensitive procedure. If one needs to test whether chemical contamination occurs above or below a defined concentration (e.g. an MRL), then clearly the method of analysis must be sufficiently sensitive to allow this to be done. However, opinions differ on how far below an MRL one should be able to measure the levels of an analyte to be confident of testing whether or not the MRL is being exceeded. Paradoxically the picture is clearer where there are no limits for the analysts to work with, when the best achievable degree of analytical sensitivity is usually the accepted norm. But this has its problems, for example in the case of genotoxic carcinogens which do not have defined no-toxicological-effect levels there is a considerable effort to keep on reducing the detection limits of analytical methods. This topic is particularly difficult to consider because the needs of providers and users of analytical data on chemical contaminants are often different.

It is unlikely that any guidelines on required analytical sensitivity would be widely accepted. However, further work needs to be done on matching the needs of those concerned, if analytical method development is to be purposeful.

11.5 BOUND OR MULTIPLE RESIDUES OF CHEMICAL CONTAMINANTS IN FOOD

Bound contaminants are those which are chemically or otherwise linked to other molecules in food. As such they are not readily extracted by the classical procedures used in food analysis which usually use organic solvents. Current methods are not generally expected to break covalent linkages between chemical contaminants and biochemical components of food. Bound chemical contaminants must be considered because they may be released after food has been eaten, for example by enzyme action in the consumer's gastrointestinal tract.

Several different chemical contaminants in a sample of food—'multiple residues'—are found from time to time and they need to be given particular attention because their combined toxicological effect, if there is one, may in theory be a synergistic one, that is greater than the sum of their individual effects. It is also possible that their combined effect is the same as or less than the sum of their individual effects.

There are few experimental data on both of these topics. Further work is needed, and to be of value extensive work would probably be required—not least because many chemical contaminants *might* be involved. At present one is restricted largely to theory, for example about the variety of different forms in which chemical contaminants might become bound to food molecules. Until some extensive practical work is carried out on these two aspects of how chemical contaminants are present in food, it will be particularly difficult to assess the degree of risk involved. Whilst it is currently thought unlikely that bound contaminants make a major contribution to dietary exposure, except in unusual circumstances, work is required to examine their significance in detail. Similarly there is some consensus that multiple chemical contamination is unlikely to be synergistic in its effects, if any, on consumers. Nevertheless, experimental data are rare because toxicology has yet to develop straightforward means of studying the effects of mixtures of chemical contaminants. Similarly analytical chemistry has yet to develop routine methods of extracting bound chemical contaminants. Research on the biological availability of chemical contaminants in the human gastrointestinal tract could provide useful models on which to base extraction methods for chemical contaminants in food.

11.6 CONCLUSIONS

Whilst much is known about chemical contaminants in food, there remains much research to be done. This chapter has summarized several areas of importance. The questions posed at the start of this chapter relate to improving the already considerable worldwide database on chemical contaminants in food. The comments and proposals that are made above are intended to stimulate a critical and thorough review of available information in this area of work and of the methods used to obtain the information. It is hoped that the thorough work done so far on chemical contaminants in food will continue to develop and expand so that a clear and extensive picture of our exposure to chemical contaminants via the diet is possible.

Index

Numbers in italics indicate where the structures of chemicals are located in the text.